Forschungs-/Entwicklungs-/Innovations-Management

Herausgegeben von
H. D. Bürgel (em.), Stuttgart, Deutschland
D. Grosse, vorm. de Pay, Freiberg, Deutschland
C. Herstatt, Hamburg, Deutschland
H. Koller, Hamburg, Deutschland
M. G. Möhrle, Bremen, Deutschland

Die Reihe stellt aus integrierter Sicht von Betriebswirtschaft und Technik Arbeitsergebnisse auf den Gebieten Forschung, Entwicklung und Innovation vor. Die einzelnen Beiträge sollen dem wissenschaftlichen Fortschritt dienen und die Forderungen der Praxis auf Umsetzbarkeit erfüllen.

Herausgegeben von

Professor Dr. Hans Dietmar Bürgel
(em.), Universität Stuttgart

Professor Dr. Hans Koller
Universität der Bundeswehr Hamburg

Professorin Dr. Diana Grosse,
vorm. de Pay, Technische Universität
Bergakademie Freiberg

Professor Dr. Martin G. Möhrle
Universität Bremen

Professor Dr. Cornelius Herstatt
Technische Universität Hamburg-Harburg

Konstantin Wellner

User Innovators in the Silver Market

An Empirical Study among Camping Tourists

With a foreword by Univ. Prof. Dr. Cornelius Herstatt

Konstantin Wellner
Hamburg, Germany

Dissertation Technische Universität Hamburg-Harburg, 2014

Forschungs-/Entwicklungs-/Innovations-Management
ISBN 978-3-658-09043-2		ISBN 978-3-658-09044-9 (eBook)
DOI 10.1007/978-3-658-09044-9

Library of Congress Control Number: 2015932768

Springer Gabler
© Springer Fachmedien Wiesbaden 2015
This work is subject to copyright. All rights are reserved by the Publisher, whether the whole or part of the material is concerned, specifically the rights of translation, reprinting, reuse of illustrations, recitation, broadcasting, reproduction on microfilms or in any other physical way, and transmission or information storage and retrieval, electronic adaptation, computer software, or by similar or dissimilar methodology now known or hereafter developed.
The use of general descriptive names, registered names, trademarks, service marks, etc. in this publication does not imply, even in the absence of a specific statement, that such names are exempt from the relevant protective laws and regulations and therefore free for general use.
The publisher, the authors and the editors are safe to assume that the advice and information in this book are believed to be true and accurate at the date of publication. Neither the publisher nor the authors or the editors give a warranty, express or implied, with respect to the material contained herein or for any errors or omissions that may have been made.

Printed on acid-free paper

Springer Gabler is a brand of Springer Fachmedien Wiesbaden
Springer Fachmedien Wiesbaden is part of Springer Science+Business Media
(www.springer.com)

Foreword

The successful development of new products requires profound knowledge about customer needs. The transfer of this knowledge is often difficult and cost-intensive ("sticky data transfer"-phenomenon), depending on the quality of the information. The resulting information gap can be completed by the application of several methods and tools: These include, among others, users who modify existing and develop completely new products ("lead user"), the application of tools to directly transfer know-how, including tacit knowledge ("user innovation toolkits"), as well as techniques of participatory observation. The mentioned phenomenon received increasing scientific attention over the past 20 years and was researched intensively. The existence and characteristics of user innovators has so far been analyzed in the areas of sports and outdoor activities, so that the focus was mainly on younger users.

At the same time, a dramatic demographic change was visible especially in industrial nations. As a result of increased life expectancies and lower birth rates, the many countries' median age of the population and especially the share of the population above 55 years are increasing. This so called "Silver Market" is growing constantly and offers assumed business opportunities for tailored products and services.

The research at hand by Mr. Wellner is the first study which analyzes the relationship between user innovation and age. The main objective of the research is the evaluation whether user innovators exist across all age groups and if yes, how older user innovators and their innovations differ from younger ones. For this, Mr. Wellner analyzes user innovations in the area of camping tourism. Methodically, he conducts only surveys in relevant communities as well as laborious on-site examinations (surveys and observations) at German camp sites. As a result, Mr. Wellner can show that older users also innovate, but differences compared to younger users are smaller than initially assumed.

The research results linked with the competent interpretation and precise presentation confirm the chosen research approach of Mr. Wellner. His essential contribution to research lies in the well-grounded discussion, application, and extention of the existing theory in the context of a relatively new phenomenon (age-based innovation). Therefore, Mr. Wellner's work constitutes an important contribution in theoretical as well as practical regards.

Hamburg, November 2014

Univ. Prof. Dr. Cornelius Herstatt

Acknowledgement

The successful development of new products requires not only technical know-how but detailed knowledge on the needs and requirements of its users. While technical know-how exists among manufacturers, knowledge on user needs resides within individuals and due to its stickiness, it is very difficult and costly to transfer. Therefore, manufacturers often develop technically sophisticated products which do not fulfill all customer needs. The remaining gap is then filled by users who modify their products or even invent new ones. This phenomenon is called user innovation and has received increasing attention by researchers during the past 20 years.

During the same time, a dramatic demographic shift – resulting from increase life expectancy and lower birth rates – has been observed, especially in industrialized countries. As a result, the average age of the population and especially the share of older people (typically defined as being above 55 years of age) has witnessed a sharp increase. The so called Silver Market is growing and therefore gaining importance. It provides opportunities for companies which offer tailored products to the Silver Market – if they understand its specific requirements.

The existence and characteristics of user innovators have been analyzed with a focus on sports and outdoor equipment which attracts especially younger individuals. My research aims to analyze relationship between age and user innovation. The key research objective is to evaluate whether user innovators also exist in the Silver Market and if so, how older user innovators and how their innovations differ from younger ones.

This dissertation would not have been possible without the ongoing help of a number of supporters, which I would like to acknowledge and thank for. Without claiming that this list is exhausting, I especially thank

- Univ. Prof. Dr. Cornelius Herstatt for convincing me to join his team and truly deserving the title "Doktorvater";
- Prof. Dr. Christan Ringle and Prof. Dr. Christian Lüthje for their ongoing feedback and advice on my research project and assuming the role of secondary evaluator and chairman of the examination committee respectively;
- all further members of the InnoAge team – Prof. Dr. Meyer, Iris, Klaus, Nils, Sandra, Nicole – for all their ideas to shape my dissertation and exploring a new field of research together;

- the TIM team, especially Carola, Daniel, Jan, Jens, Moritz, Niklas, Nils, Raj, Sarah, Stephan, Thorsten, Tim, Tim, and Viktoria for all the fruitful discussions, brainstorming sessions over a cup of coffee, and all the fun times;
- the Forschungs- & Wissenschaftsstiftung Hamburg, e-fellows.net, and the Firm for supporting me both financially and intellectually;
- the participants of several conferences and doctoral seminars for all the valuable input and challenging questions;
- the adminstrators and members of campen.de, camperfreunde.com, camperboard.de, ClassiCaravan, klappcaravanforum.de, and wohnwagenforum.de, as well as the hundreds of respondents on several German camp sites who all invested their valuable time to support my research;
- Andreas and Tina for sharing their (frustrating) PhD moments – your development was (and still is) inspiring;
- My family for always providing a safe haven and and putting trust in me.

Most of all, I thank Verena not only for accompanying me on this journey, but for being the best travel companion one could ever imagine and being a true soulmate.

Munich, December 2014

Konstantin Wellner

Table of Contents

Foreword V
Acknowledgement VII
Table of Contents IX
Index of Figures XIII
Index of Tables XV
List of Abbreviations and Symbols XVII

1 Introduction 1
   1.1   Motivation and Research Objectives 1
   1.2   Research Approach and Contributions 3
   1.3   Structure of the Dissertation 5
   1.4   Key Definitions 6

PART A. THEORETICAL FOUNDATION 9

2 The Silver Market Phenomenon 9
   2.1   Demographic Development and Transition 9
   2.2   Silver Market Phenomenon 13
         2.2.1 *Description of Silver Market* 14
         2.2.2 *Product Development for the Silver Market* 16
   2.3   Defining Age 17
         2.3.1 *Shortcomings of Chronological Age and Alternative Age Measurements* 18
         2.3.2 *Cognitive Age* 19
   2.4   Effects of Aging 20
   2.5   Age and Innovative Behavior 22
   2.6   Interim Conclusions 24

3 Users as Main Source of Innovations 26
   3.1   Development of User Innovation Research 26
   3.2   Users as the Main Functional Source of Innovation 27
   3.3   Characteristics of User Innovators 34
   3.4   Lead User Theory 35
   3.5   Antecedents of Lead User Characteristics and Innovative Behavior 38
   3.6   Development of User Innovation in Academic Research 40
   3.7   Interim Conclusions 42

4 Research Questions and Hypotheses 43
4.1 Research Gap and Research Questions 43
4.2 Hypotheses Regarding Silver Market User Innovators 44
4.2.1 Use Experience 44
4.2.2 Product Knowledge 45
4.2.3 Technical Expertise 46
4.2.4 Lead User Characteristics 48
4.2.5 Moderating Influence of Age 49
4.3 Propositions Regarding Innovation Characteristics of Silver Market User Innovators 54

PART B. QUANTITATIVE EMPIRICAL STUDY 61

5 Introduction to the Research Field: Camping & Caravanning 61
5.1 Characterization of Camping Market 61
5.1.1 Origin and History of Camping 61
5.1.2 Camping in Germany and around the World 64
5.2 Reasons for Selection of Camping & Caravanning Industry 67

6 Explorative Survey among Companies 69
6.1 Motivation for Study and Selection of Questions 69
6.2 Selection of Companies 69
6.3 Results of Company Survey 70
6.4 Interim Conclusions 74

7 Empirical Study among Camping & Caravanning Tourists 75
7.1 Research Design and Operationalization 75
7.1.1 Structural Equation Modeling with PLS 75
7.1.2 Operationalization of Constructs 79
7.1.3 Data Collection and Sample Description 87
7.1.4 Data Cleansing and Preparation 92
7.2 Findings Regarding Silver Market User Innovators 94
7.2.1 Results of Descriptive Analysis of Survey Results 94
7.2.2 Findings Regarding Correlations of Chronological Age and Cognitive Age 99
7.2.3 Existence of User Innovators across Age Groups ... 103
7.2.4 Statistical Tests and Bias Treatment 107
7.2.5 Model Evaluation 111
7.2.6 Measurement Model 112
7.2.7 Evaluation of Structural Model – Determinants of Innovative Behavior 125
7.2.8 Mediator Analysis for High Expected Benefits 129
7.2.9 Evaluation of Control Variables 130
7.2.10 Interaction Effect of Age on Structural Model 131
7.2.11 Testing for Non-Linear Effects from Use Experience ... 137
7.2.12 Characterization of Silver Market User Innovators and Non-Innovators 138
7.2.13 Evaluation of Hypotheses 140

7.3 Findings Regarding Silver Market User Innovations and Related Processes 141
    7.3.1 Descriptive Analysis of Survey Results 141
    7.3.2 Impact of Motivational Factors on the Innovation Characteristics 146
    7.3.3 Impact of Age on Innovation Characteristics 149

PART C. DISCUSSION & CONCLUSIONS 153

8 Discussion 153

8.1 RQ1: Do User Innovators Exist in the Silver Market Population? 153
8.2 RQ2: Which Determinants of Innovative Behavior Characterize the Silver Market User Innovator? Do these Determinants Differ Compared to Younger User Innovators? 154
    8.2.1 General Determinants of Innovative Behavior as a Baseline 155
    8.2.2 Differences on Determinants of Innovative Behavior between Silver Market and Non-Silver Market User Innovators 156
    8.2.3 Comparison of Demographic Characteristics of Innovators and Non-Innovators in the Silver Market Population 161
    8.2.4 Summary and Response to Research Question 162
8.3 RQ3: How Strong - If There Is One - Is the Moderating Influence of Chronological/Cognitive Age on the Determinants of Innovative Behavior? 163
8.4 RQ4: Do User Innovations by Silver Market User Innovators Differ from "Regular" User Innovations, and if so, How? 164

9 Contribution and Implications 170

9.1 Contributions to Academic Research 170
    9.1.1 Implications for Innovation Management 170
    9.1.2 Implications for Silver Market Theory 171
    9.1.3 Implications for Measuring Age in Innovations Research 172
9.2 Recommendations for Managerial Practice 173
9.3 Limitations and Suggestions for Further Research 176

References 179
Appendix 204

Index of Figures

Figure 1: Development of Fertility Rate and Life Expectancy from 1950 - 2050 10
Figure 2: The Demographic Transition and Population Growth Rate over Time 10
Figure 3: Development of Older Population from 1950 - 2050 12
Figure 4: Development of Age Structure in Germany from 1990 - 2030 13
Figure 5: Number of Thomas Edison's US Patents by Age, based on Execution Date 24
Figure 6: Development of Scientific Articles on User Innovation in Peer-Reviewed Publications from 1959 to 2012 41
Figure 7: Overview of Hypotheses Regarding Silver Market User Innovators without Moderating Influence of Age 48
Figure 8: Dimensions of Innovation Characteristics 55
Figure 9: Age Distribution of German Campers versus Non-Campers 66
Figure 10: Selection of Approached Camping Companies 70
Figure 11: Participation with Customers during Ideation and Product Development . 71
Figure 12: Evaluation of Customer Ideas and Prototypes 72
Figure 13: Reasons for Not Realizing Customer Ideas 73
Figure 14: Structural Equation Model with Latent Variables 76
Figure 15: Online Camping Communities in Germany (as of February 24th, 2014) ... 89
Figure 16: Overview of Approached Campsites in Germany 91
Figure 17: Distribution of Age 94
Figure 18: Distribution of Income 95
Figure 19: Use Experience Box Plots across Age Groups 96
Figure 20: Innovator Share of Disposable Time 97
Figure 21: Technical Expertise of Innovators versus Non-Innovators 99
Figure 22: Distribution, Scatter Plots, and Correlations of Age Constructs 100
Figure 23: Comparison of Age Difference and Chronological Age 101
Figure 24: Distribution of Age Differences between Innovators and Non-Innovators above 55 Years 103
Figure 25: Innovator Shares across Age Groups 104
Figure 26: PLS Model for Evaluation of Common Method Bias 110
Figure 27: Results of Structural Model 1 127
Figure 28: General Mediator Model 129

Figure 29: Transcription of Structural Model with Moderator for PLS Path Modeling 132
Figure 30: Furthest Development Stage of Innovations 143
Figure 31: User Self-Classifications of Innovations 144
Figure 32: Comparison of Absolute Mean Differences with Different Separators of Chronological Age 149
Figure 33: Development Time of User Innovators 211
Figure 34: Development Frequency of User Innovators 211
Figure 35: Cooperation during Ideation and Realization Phase 211

Index of Tables

Table 1:	Selected Studies on User Innovations	32
Table 2:	Overview of Studies Analysing Influencing Factors of Innovative Behavior and Lead User Components	39
Table 3:	Expert Evaluations of Moderating Impact of Age	50
Table 4:	Sociodemographic Characteristics of Caravan Owners Compared to the General Population in Germany	67
Table 5:	Applied Tools for Customer Integration	73
Table 6:	Operationalization of Constructs	80
Table 7:	Evaluation of the Self-Completion Questionnaire in Relation to the Structured Interview	88
Table 8:	Responses from Online Survey	90
Table 9:	Responses of Paper-based Survey	92
Table 10:	Disposable Time and Innovator Share of Occupation Status	97
Table 11:	Characteristics of Total Sample, Innovators, and Non-Innovators	98
Table 12:	Comparison of Silver Market Shares of Innovators Considering the Adjusted Age at Innovation	105
Table 13:	Evaluation of Common Method Bias	111
Table 14:	Results of Exploratory Factor Analysis for Lead Userness	116
Table 15:	Results of Initial Exploratory Factor Analysis for Ahead of Trend	116
Table 16:	Results of Exploratory Factor Analysis for High Exptected Benefits	116
Table 17:	Results of Initial Exploratory Factor Analysis for Technical Expertise	116
Table 18:	Results of Exploratory Factor Analysis for Product Knowledge	117
Table 19:	Results of Exploratory Factor Analysis across all Selected Items	117
Table 20:	Final Results of EFA for all Reflective Constructs (measured separately)	118
Table 21:	Evaluation of All Items of Reflective Constructs of Measurement Model	121
Table 22:	Evaluation of Selected Items of Reflective Constructs of Measurement Model	122
Table 23:	Cross Loadings	122
Table 24:	Discriminant Validity (Fornell-Larcker Criterion)	122
Table 25:	Evaluation of Formative Measure "Use Experience"	125
Table 26:	Quality Criteria of Structural Model 1	128
Table 27:	Results of Interaction Model 1 and 2	133
Table 28:	Results of Multi-Group Analysis Regarding Age	135

Table 29:	Comparison of Characteristics of Innovators and Non-Innovators in the Silver Market Segment	139
Table 30:	Evaluation of Hypotheses	141
Table 31:	User Self-Ratings of Innovation Qualities	146
Table 32:	Correlation Coefficients of Motivators with Process Qualities, Innovation Types, and Innovation Qualities	147
Table 33:	Summary of Findings Regarding Mean Differences with Respect to Different Age Measurements	150
Table 34:	Summary of Findings Regarding Mean Differences with Respect to Age Differences of Older User Innovators	151
Table 35:	Results of Test for Mode Effects on Measurement	205
Table 36:	Indicator and Construct Reliability	206
Table 37:	PLS Cross-Loadings	206
Table 38:	Outer Loadings, Weights, and Multicollinearity of Formative Constructs	206
Table 39:	Indicator and Construct Reliability	207
Table 40:	PLS Cross-Loadings	207
Table 41:	Outer Loadings, Weights, and Multicollinearity of Formative Constructs	207
Table 42:	PLS-MGA for FEEL Age and LOOK Age Groups	208
Table 43:	PLS-MGA for DO Age and INTEREST Age Groups	208
Table 44:	PLS-MGA for Age Difference Groups in the Full and the Silver Market Sample	209
Table 45:	Results of Main Effects Model with Control Variables	210
Table 46:	Correlation Coefficients of Age Measurements with Innovation Characteristics	212

List of Abbreviations and Symbols

Ø	Average
%	Percent
ADAC	Allgemeiner Deutscher Automobil-Club e.V. (eng.: General German Automobile Association)
AoT	Ahead of Trend
Asympt.	Asymptotic
B2B	Business-to-Business
BVCD	Bundesverband der Campingwirtschaft in Deutschland e.V. (eng.: Federal Camping Association of Germany)
CB-SEM	Covariance-based Structural Equation Modeling
CEO	Chief Executive Officer
cf.	Confer
CFA	Confirmatory Factor Analysis
CITC	Corrected Item-to-Total-Correlation
CIVD	Caravaning Industrie Verband e.V.
d	Day(s)
e.g.	Exempli gratia (eng.: for example)
eng.	English
EFA	Exploratory Factor Analysis
ELU	Embedded Lead User
HEB	High Expected Benefits
eng	English
et al.	Et alii (eng.: and others)
etc.	Et cetera (eng.: and so forth)
f^2	Effect Size
EUR	Euro
GDR	German Democratic Republic
GoF	Goodness-of-Fit
h	Hour(s)
H_0	Null Hypothesis
i.e.	Id est (eng.: that means)
IB	Innovative Behavior
IIC	Inter-Item-Correlation
Inc.	Income
IP	Intellectual Property
ISSN	International Standard Serial Number
ISCO	International Standard Classification of Occupations
KMO	Kaiser-Meyer-Olkin
LES	Leading Edge Status
LU	Lead Userness
LV	Latent Variable
MAR	Missing at Random
MCAR	Missing Completely at Random
MGA	Multi-Group Analysis
MNAR	Missing Not at Random

MSA	Measure of Sampling Adequacy
N	Sample Size
n/a	Not Applicable
n.s.	Not Significant
OD	Omission Distance
OECD	Organisation for Economic Co-operation and Development
p	Probability
PLS	Partial Least Squares
PK	Product-related Knowledge
Q^2	Predictive Relevance (measured with the Stone-Geisser Q^2)
r	Correlation coefficient
R^2	Coefficient of Determination / Explained Variance
R&D	Research and Development
RQ	Research Question
SD	Standard Deviation
SEM	Structural Equation Model / Structural Equation Modeling
SiA	Silver Age(r)
Sig.	Significance
SiMa	Silver Market
SME	Small and Medium Enterprises
TE	Technical Expertise
TÜV	Technischer Überwachungsverein (eng.: Technical Inspection Association)
UE	Use Experience
UK	United Kingdom
US	United States
y	Year(s)

1 Introduction

1.1 Motivation and Research Objectives

Over the last decades, the world's population and its structure have changed quickly. The global population grew from 3 billion people in 1960 to over 7 billion people now.[1] The median age of the world's population increased from 23 years in 1960 to 29 years today and is expected to grow to 36 years in 2050. This change is even stronger and faster in industrialized countries. Germany and Japan are currently among the oldest nations in the world. Their median age grew from 1960 until now from 35 years to 44 years (Germany) and from 26 years to 45 years (Japan).[2] While the baby-boomer generation[3] fueled economic growth and prosperity, their transition to the retirement age creates problems. The resulting shortage of the labor force and growing challenges for the pension and care systems create social and intergenerational tensions, especially against the background of financial and economic uncertainty. But the demographic change does not merely present a threat; it also provides business opportunities. Today's elderly demand products that fulfill their requirements for quality, comfort, and security, while helping them to continue to lead an active and autonomous lifestyle.[4] Aging also negatively affects the physical, sensory, and cognitive capabilities. Products that are designed for younger users might therefore not be suitable for older users anymore. The market for specific age-based innovations becomes more attractive because the number of potential customers above 55 years is constantly growing. This market is typically called the "Silver Market" (SiMa).[5] Kohlbacher and Herstatt (2011b) state about this market segment: *"Increasing in number and share of the total population while at the same time being relatively well-off, this market segment can be seen as very attractive and promising, although still very underdeveloped in terms of product and service offerings."*[6] Although the attractiveness of the SiMa has been realized, many companies do not specifically target it, and the integration of users in the innovation process is still hallmarked by "[...] *numerous unrealized opportunities [...]. One*

---

[1] Cf. United States Census Bureau 2013.
[2] Cf. United Nations 2013.
[3] The generation born after World War II, i.e., between 1945 and 1965, is typically referred to as baby boomers.
[4] Cf. Arnold & Krancioch 2011, p. 155; Usui 2008, pp. 73 & 334; Reinmöller 2008, p. 160; Tempest et al. 2008, p. 247.
[5] See Kohlbacher & Herstatt 2008b; Hedrick-Wong 2007; Kunisch et al. 2011.
[6] Kohlbacher & Herstatt 2011b, p. vii.

possible reason for this exclusion is that there is a lack of valid and reliable empirical research available to help guide marketing strategies."[7]

If older users have different demands compared to younger ones but these are not well represented in manufacturers' product development processes, how else can age-based products be created? At this point, the analysis of user innovations and the application of the lead user method[8] provide valuable insights into the specific demands and corresponding solutions of users, currently not served by the market.

Before Eric von Hippel discovered in 1976 *"[...] that the innovators are most often users"*[9] the generally accepted belief was that manufacturers are solely responsible for the development of products and the whole innovation process. More than twenty years later in 1998, Steve Jobs told Businessweek: *"A lot of times, people don't know what they want until you show it to them."*[10] Although he was beyond doubt a brilliant manager and innovator, Jobs was only partly right with this assessment. Product innovation failure rates across industries are assessed to be between 40 % and 90 %;[11] for fast-moving consumer goods they can even be 70 % to 90 %[12]. While some of the reasons accounting for this failure are product-based, in that some manufacturers do not offer a compelling advantage over existing products,[13] other reasons pertain to the manufacturer's insufficient need knowledge and developer overconfidence[14]. The integration of users in the innovation process can reduce failure rates. For this purpose, several methods to assist companies were developed, e.g., the lead user method, innovation communities, and toolkits,[15] which all have a positive impact on the success of innovations[16]. Lead users, for example, have specific needs long time before the general market.[17] Since the existing market offering often does not suit their needs, they develop solutions on their own and, therefore, they indicate market trends and create innovations. The existence of user innovators does not only provide benefits for companies but also for society and the economy as a whole. User innovators develop solutions for markets whose demands are not large enough for commercial offerings. Therefore, they provide products for

---

[7] Sudbury & Simcock 2009, p. 23.
[8] Cf. Hippel 1986, pp. 797ff.; Urban & Hippel 1988.
[9] Hippel 1988, p. 11.
[10] Reinhardt 1998.
[11] Cf. Cierpicki et al. 2000, p. 777; Crawford 1977, p. 51; Griffin 1997, pp. 431f. & 438.
[12] Cf. Gourville 2005, p. 5.
[13] Cf. Rogers 2003, pp. 229ff.
[14] Cf. Gourville 2005, p. 7.
[15] See Hippel 1986; Herstatt & Hippel 1992; Hippel 2005a, pp. 93ff.; Franke & Hippel 2003.
[16] Cf. Franke et al. 2006, pp. 310ff.; Hippel et al. 1999, p. 56; Herstatt & Hippel 1992, p. 219.
[17] Cf. Hippel 1986, p. 791.

unserved market niches and improve the overall economic wealth.[18] The newest findings suggest that user innovators do not merely exist in small special-interest market niches. Rather they are actually a mass market phenomenon. The investments of user innovators can even outnumber the commercial R&D spending of whole countries.[19]

Research on the behavioral, psychological, biological, and societal aspects of aging is combined in the field of gerontology. The implications of its findings are mostly formulated for policy makers and whole national economies and can only rarely provide recommendations for managerial practice.[20] Especially innovative behavior has to date only been researched within the boundaries of organizational and human resource management. Therefore, Astor (2000, p. 322) called for research in innovation management on the impact of age on innovation. Findings on creative output over one's life course indicate that user innovators exist among the elderly and that they can therefore be integrated into manufacturers' innovation processes.[21] The commonalities and differences that exist between older and younger user innovators are currently unknown. Therefore, the methods to integrate users in product development and diffusion cannot be adapted to cater to the SiMa.

Hence, the research objective of this study is to evaluate whether there exist user innovators in the SiMa. If they exist, the further aim is to determine how they and their innovations might differ in order to provide academics and managers with insights on how to integrate them best in the innovation process.

1.2 Research Approach and Contributions

To respond to the research objective and the specific research questions (see chapter 4.1 below), an empirical approach was applied. Since the existence of lead users has been proven for several cases and empirically derived findings on user innovators in younger age groups is already available, an empirical approach allows for the testing of hypotheses.[22]

---

[18] Cf. Harhoff et al. 2003, p. 1768.
[19] Cf. Flowers et al. 2010, p. 15; Hippel et al. 2012, p. 1675; Hippel et al. 2011, p. 28.
[20] The largest research institute in Germany, the German Centre for Gerontology (Deutsches Zentrum für Altersfragen) for example emphasizes in its research statement that it *"[...] participates in the provision of knowledge on age-related issues and supplies society, politicians and the academic debate with up-to-date, innovative information on age and ageing issues."* http://www.dza.de/en/research.html, accessed on January 27, 2014.
[21] See Simonton 1988.
[22] Cf. Bortz & Döring 2009, p. 52.

Since the needs, motivations, and determinants of behavior of individuals are in the focus of the research, the bulk of this thesis is based on a survey conducted among individual users. The survey was conducted among camping and caravanning tourists because all age groups are represented in this product field and the product characteristics allow for user innovations (see chapter 5.2 for a detailed explanation). The survey was conducted in six German online camping communities and on nine German campsites. In total 351 usable responses were collected. Among these were 157 users with innovative ideas and 103 who had even developed a working prototype. The most important antecedents for lead userness and innovative behavior – use experience, product knowledge, and technical expertise[23] – were derived from the literature and analyzed with partial least squares structural equation modeling (this included a multi-group analysis for the selected age groups). Besides the traditional chronological age, cognitive age was also applied as an alternative age measurement to test its applicability for the segmentation of user innovators in the SiMa and the prediction of their behavior.

The findings of this study contribute to existing research in innovation management as well as to gerontology. This is the first study, which explicitly investigates the relationship between user innovation and age. The age range of the sample ranges from 19 to 86 years. One can therefore compare the innovative behavior of younger and older age groups and is not limited to individuals who are not yet retired, as is always the case in organizational research. The findings show that user innovators do not only exist in a very specialized product environment but also in a low-tech field with many participants. This confirms the first findings of Hippel, Ogawa, and Jong (2011) and Hippel, Jong, and Flowers (2012), that user innovations are a mass market phenomenon.

This study shows that the importance of the antecedents for innovative behavior, as well as the independence of the two components of lead userness, change with increasing age. While a general predisposition to be an innovator may remain constant over time,[24] the relative importance of an individual's knowledge and resources that facilitate innovative behavior change with age. This study also uncovers that not all discussed antecedents necessarily have a linear effect. Non-linear effects have not yet been discussed in the literature, and more careful analysis and interpretation is needed – especially in the cases of older people for whom the differences in experiences and knowledge are largest.

---

[23] See Schreier & Prügl 2008; Franke et al. 2006; Lüthje 2004; Slaughter 1993; Lüthje 2000; Tietz et al. 2005.

[24] Cf. Rogers 2003, pp. 248 & 269f.

The study further contributes to an understanding of which age measures are most suitable for innovation management. While chronological age is traditionally used, its explanatory power is limited because not all significant changes in life are tied to chronological age. While the chronological age determines many formal aspects of the life course (e.g., the start of education around the age of 6, retirement around the age of 65), other determinants that influence behavior, like health status, marital status, and the social network, are not caused by aging alone.[25] Therefore, not all 60- or 70-year-olds are the same, and some researchers even argue that the heterogeneity within the SiMa is at least as large as the one between age cohorts.[26] In this context, the applicability of Barak's (2009) cognitive age as an alternative age measure in innovation research is analyzed. The resulting age difference from cognitive age and chronological age is often only used as a descriptive statistic. The age difference is interpreted in relation to the innovative behavior of the individuals which provides valuable insights.

1.3 Structure of the Dissertation

This dissertation consists of nine chapters in three parts. This first chapter provides an introduction to the research area and the applied approach. Part A of the dissertation consists of three chapters and presents the theoretical foundations. Chapter 2 defines the Silver Market phenomenon and highlights the demographic shift and its effect on markets and product development. The shortcomings of market segmentation based on chronological age are explained and the biological, social, and psychological effects of aging are outlined. Chapter 3 deals with users as the main source of innovations, providing an overview on current findings and specifically focusing on lead user theory. In chapter 1, the research questions and the resulting hypotheses on the determinants of innovative behavior and the influence of age are derived.

Part B begins with an introduction to the research field of camping and caravanning, which is used for the empirical part of the dissertation, in chapter 5. Chapter 1 explores how innovations by users are evaluated by companies in the research field and whether users are actively integrated in the development process. Chapter 1 contains the main study of this dissertation. The operationalization of the theoretical constructs is followed by a description of structural equation modeling with partial least squares. The majority of the chapter presents the results of the empirical

---

[25] Cf. Super 1994, p. 254.
[26] Cf. Sudbury & Simcock 2009, p. 32; Dannefer 1987, pp. 228f.

analysis regarding the attributes of user innovators, determinants of innovative behavior, and the characteristics of the resulting innovations.

Part C consists of two chapters. Chapter 8 discusses the findings and compares them to previous research. The final chapter, 1, summarizes the contributions and highlights implications for academic research and recommendations for managerial practices. Finally, the limitations of this study are specified and suggestions for further research are provided.

1.4 Key Definitions

This chapter will provide definitions for some of the key terms used throughout this dissertation. The concepts will be further detailed in their respective chapters.

Innovation: The OECD defines innovation as "*[...] the implementation of a new or significantly improved product (good or service), or process, a new marketing method, or a new organisational method in business practices, workplace organisation or external relations.*"[27] This definition shows that innovations occur in different settings. For the consumer goods market, product innovations are most relevant. A product innovation is specifically defined as "*[...] the introduction of a good or service that is new or significantly improved with respect to its characteristics or intended uses. This includes significant improvements in technical specifications, components and materials, incorporated software, user friendliness or other functional characteristics.*"[28] Some scholars also separate invention from innovation. While an invention is the sole creative act of developing something new, an innovation also includes the commercialization of the product.[29] In this research project, this differentiation is not made. The improvement of products in active use would be implementation enough.[30] Commercialization is mostly not suitable for private users. In this project, the minimum requirement for innovation is the clear articulation of an improvement idea. A restriction to provide at least a working prototype would be too narrow in the case of consumer innovations because implementation might be confined by individually limited resources of time, money or technical capabilities.[31]

---

[27] Organisation for Economic Co-operation and Development (OECD) & Statistical Office of the European Communities (Eurostat) 2005, p. 46.
[28] Organisation for Economic Co-operation and Development (OECD) & Statistical Office of the European Communities (Eurostat) 2005, p. 48.
[29] Cf. Ogawa & Pongtanalert 2013, p. 43.
[30] Cf. Slaughter 1993, p. 85.
[31] Cf. Ernst et al. 2004, pp. 125f.

Key Definitions

Innovator: In the context of this study, an individual who developed an innovation as described above is an innovator. Since this study is not interested in the innovator of specific products, but rather in the innovative behavior of people, it is not necessary for the innovator to be the first or only person to develop the innovation, as long as they do not know about any similar innovations.

User innovator: If, at the time of the innovation, the innovator expected to benefit solely from *using* the product, it is classified a user innovation.[32] In contrast, manufacturers benefit from *selling* a product or service. According to this definition, individuals and firms can be user innovators.

Lead user: Lead users are users who fulfill the following two criteria, defined by Hippel (1986, p. 796) [emphasis is original]:

- *"Lead users face needs that will be general in a marketplace-but face them months or years before the bulk of that marketplace encounters them,* **and**
- *Lead users are positioned to benefit significantly by obtaining a solution to those needs."*

It is important to note that lead users are not necessarily user innovators or early adopters. Rather, due to their exposed position, they are more likely to develop their own solutions or to belong to the first adopters once a commercial product is available.[33]

Silver Market: The Silver Market (SiMa) refers to the market segment of older consumers. 55 years or older is the typical minimum age to be considered an older consumer and this threshold is also applied in this study.[34] Members of the Silver Market are also called Silver Agers (SiA).

Age: During the course of this study, age refers to chronological age which is measured in full years since the date of birth.

Age-based innovations: Age-based innovations (or products) are specifically designed with the needs and requirements of older consumers in mind. These products can either be adaptations of existing products, new products designed specifically for the SiMa or age-neutral products with a universal design.[35] Products

---

[32] Cf. Hippel 2005a, p. 3, 2005b, p. 64, 2007, p. 294; Shah 2000, p. 7.
[33] Cf. Hippel 2007, p. 300.
[34] Cf. Auken et al. 2006, p. 440; Fisk et al. 2009, p. 8; Moschis 1992b, p. 21. A more detailed description is provided in chapter 2.2.
[35] Cf. Kohlbacher et al. 2011b, p. 5; Iffländer et al. 2012, p. 11; Gassmann & Reepmeyer 2006, p. 124.

that are merely separately marketed for the SiMa are not considered to be age-based innovations in this study.

Part A. THEORETICAL FOUNDATION

2 The Silver Market Phenomenon

2.1 Demographic Development and Transition

The number of humans living on Earth has been constantly increasing. While the global human population did not reach 1 billion until the beginning of the 19th century, it has been growing rapidly ever since. The population reached two billion in 1927 (123 years later), 3 billion in 1960 (33 years later), and 4 billion in 1974 (14 years later). Since then an additional billion has been added approximately every 12 to 14 years, culminating in over 7 billion people today.[36]

The key drivers of population size and growth are mortality, fertility, and migration, but migration is irrelevant on a global level. Improvements in medicine and healthcare, e.g., discovery of penicillin, nationwide immunizations, and precautions against communicable diseases, have led to significantly lower mortality rates and higher life expectancies across all regions (see Figure 1 below). The world's life expectancy at birth increased from 46.9 years in 1950-55 to 70.0 years in 2010-2015. The current life expectancy for the most developed countries is even higher: 78.9 years in the US, 80.7 years in Germany, and 83.5 years in Japan.

Fertility is measured in accordance with the fertility rate, which is defined as *"[...] the average number of children a woman would bear over the course of her lifetime [...]"*[37]. A fertility rate of 2.1 is required for constant reproduction. Over the past 50 years fertility rates have constantly declined (see Figure 1 below). While they were already well below the reproduction rate in developed countries, the less and least developed countries have seen an especially sharp decline. The decreasing fertility rate leads to a slower overall population growth. While the annual population growth rate peaked in the late 1960s at 2.1 %, it is currently at 1.3 % and will continue to decline. Nevertheless, the global population will continue to grow during the 21st century and is expected to stabilize at just above 10 billion people after 2200.[38]

---

[36] Cf. United Nations 1999, p. 8. Several studies have tried to estimate historical population figures. Two good overviews on existing studies and their key findings can be found under the following links:
http://en.wikipedia.org/wiki/World_population_estimates
http://www.census.gov/population/international/data/worldpop/table_history.php
[37] United Nations 2010, p. 60.
[38] Cf. United Nations 2013.

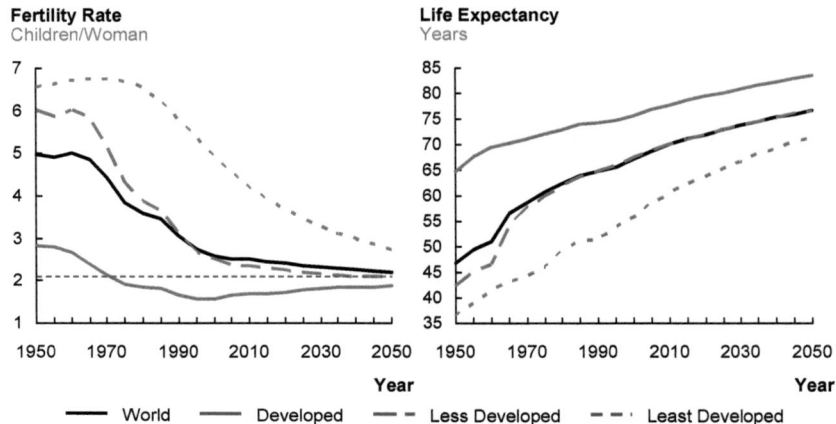

Figure 1: Development of Fertility Rate and Life Expectancy from 1950 - 2050[39]

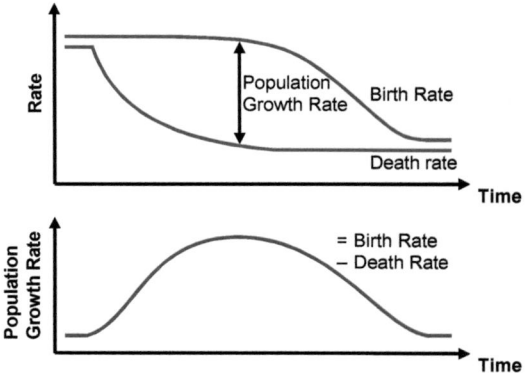

Figure 2: The Demographic Transition and Population Growth Rate over Time[40]

The decrease in fertility typically begins only after the decrease in mortality has already become apparent. This lag creates population growth (see Figure 2 above)

---

[39] Own illustration. Data based on United Nations 2013, medium-variant scenario.
Developed countries comprise of Europe, North America, Australia, Japan, and New Zealand.
Less developed countries comprise of Africa, Asia (excluding Japan), Latin America, the Caribbean, Melanesia, Micronesia, and Polynesia without the least developed countries.
Least developed countries comprise of 48 countries (33 in Africa, 9 in Asia, 1 in Latin America, and 5 in Oceania) as designated by the United Nations General Assembly in 2011.

[40] Illustration according to Bloom et al. 2003, p. 31.

and triggers a demographic transition.[41] At first a large cohort is born and, as it traverses through the working ages, it fuels economic growth and provides a demographic dividend.[42] This large cohort of the current demographic transition is the so called *baby boomer generation* born between 1940 and 1960. Once that large cohort is past the median age of the population, the older age cohorts of a population grow at a higher rate than the average population, leading to a demographic burden. The resulting phenomenon is the overall aging of the population. This is visible as an increasing median age and a growing share of older age cohorts. The median age of the world population grew from 23.5 years in 1950 to 28.5 years in 2010. In Japan, currently the oldest nation in the world, the median age more than doubled in the same time period from 22.3 years in 1950 to 44.9 years in 2010. The older segment of the population (aged 60 years or over) currently accounts for about 11 % of the global population, but its share is expected to increase to 22 % (over 2 billion people) by 2050.[43] In Japan and Germany, this segment already accounts for more than 31 % and 26 %, respectively (see Figure 3 below). While most of the growth of the older age cohorts in recent years has come from developed countries, in the future it will be driven by growth in the less developed regions of Africa, Asia, and Latin America.[44]

As stated above, Germany (alongside Japan) has already experienced the demographic transition and is currently one of the oldest countries. Its fertility rate is currently at 1.42 (recovering from an all-time low after the reunification at 1.30), and life expectancy at birth is currently 78.2 years for men and 83.1 years for women. The additional life expectancy at the age of 60 was 22 years for men and 25 years for women, which means that a German man at the age of 60 today will on average live until he is 82. As a result, the median age grew from 35.3 years in 1950 to 44.3 years in 2010 and is expected to rise even further to 51.5 years in 2050.[45]

Based on data by the Federal Statistics Office of Germany, there are currently 28 million people of at least 55 years of age living in Germany. They account for 35 % of the overall population.[46] Since 1990, the share of that age group has grown from 22 % and is expected to reach 42 % in 2030 (see Figure 4 below).[47] The shape of Germany's population age structure will then change from a pyramid to something like a mushroom.

---

[41] Cf. Bloom et al. 2003, pp. 30ff..
[42] Cf. Fent et al. 2008, pp. 4f.
[43] Cf. United Nations 2012, p. 1.
[44] Cf. United Nations 2012, p. 1.
[45] Cf. United Nations 2013; Statistisches Bundesamt Deutschland 2011, pp. 10ff.
[46] The explanation for defining the cut-off value at 55 years will be delivered in chapter 2.2.
[47] Cf. Statistisches Bundesamt Deutschland 2009.

Median Age

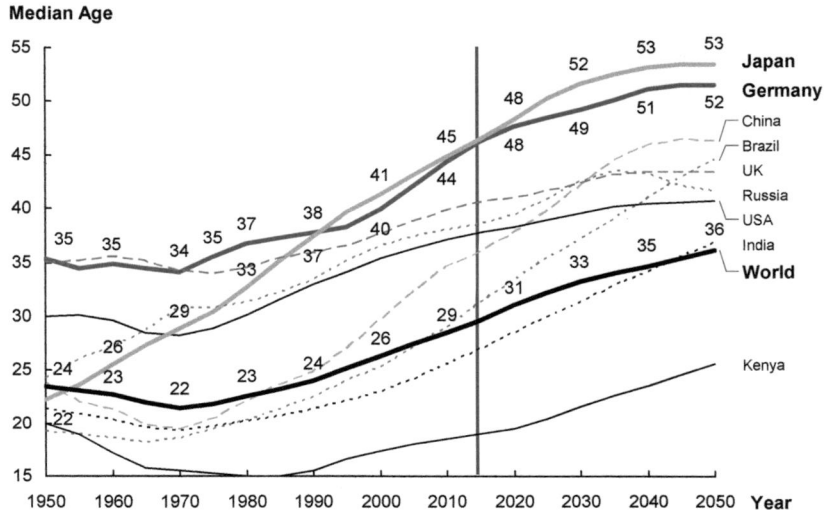

Figure 3: Development of Older Population from 1950 - 2050[48]

Along with the changing age structure also come challenges for the social systems of these countries, as the (decreasing) members of the workforce must support the (increasing) transfer recipients. Common measures to express the amount of pressure that is on the workforce are dependency ratios or, more specifically, old-age dependency ratios. The old-age dependency ratio is defined by the ratio of people aged 65 and older and the number of people within the age limits of the workforce (15 - 64) represented as the number of dependents per 100 persons of working age. Although there is considerable criticism regarding the simplifying assumptions of this measure, it is still commonly applied.[49] The old-age dependency ratio in Germany is currently at 33 and is estimated to increase to 60 by 2050. In other words, one person 65 or older is currently supported by 3.1 members of the workforce. In 2050, this ratio will be reduced to only 1.7. For Japan this ratio will drop from 2.4 (the current ratio) to 1.4 in 2050. Globally, the old-age dependency ratio is currently at 13

---

[48] Own illustration. Data based on United Nations 2013, medium-variant scenario.

[49] Typically criticism concerns the notion that it only compares the sizes of the age groups without incorporating the fact that some old people might still be members of the workforce while some middle-aged people may not. Additionally, the value of transfers is not included. The measure assumes that the cost for supporting a child and supporting a retired person is equal. More fundamental critics argue that the term dependency ratio already implies that population aging is a burden to society and neglects the idea that older people are the source for many financial transfers to younger generations, especially in developed countries. For an overview, the reader may refer to Crown 1985.

(≙ 7.9 supporting workforce members) and is estimated to increase to 25 (≙ 4.0 supporting workforce members).

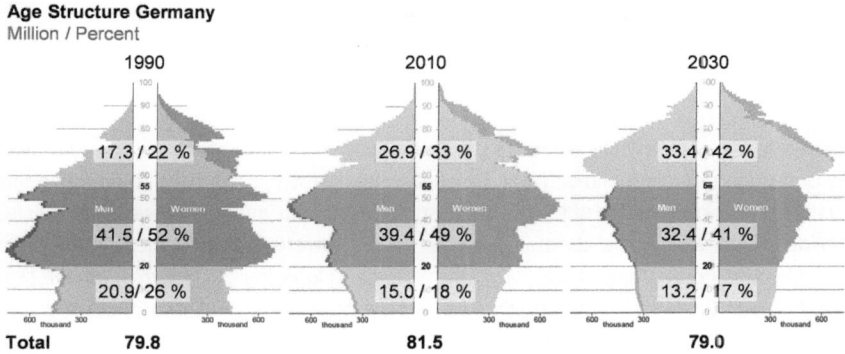

Figure 4: Development of Age Structure in Germany from 1990 - 2030[50]

As shown, the demographic transition, especially in developed countries, presents societies with tremendous challenges. Social support systems must accommodate an increasing number of the elderly, and many states do not even have a public pension system in place.[51] But the demographic transition is not solely a threat. The soon-to-retire baby boomers are well educated, healthy, and wealthy and can be a great business opportunity for tailored products, as the following chapter will discuss.

2.2 Silver Market Phenomenon

There are currently around 800 million people in the world who are 60 years or older. About a third of them are living in developed countries. By 2050, there will be more than 2 billion people of at least 60 years of age, and most of this growth, about 80 %, or 1.6 billion people, will come from developing countries.[52] The sheer size and rapid growth of this age group, coupled with the assumption that the group has different needs than younger age groups provide promising business opportunities for tailored products and services.

---

[50] Own illustration based on Statistisches Bundesamt Deutschland 2009. Model 1-W2 with the following assumptions: nearly constant birth rate at 1.4 children per woman, life expectancy of newborns in 2060 at 85.0 years for boys and 89.2 years for girls and a positive annual net migration of 200,000 persons.
[51] Cf. United Nations 2012, p. 4.
[52] Cf. United Nations 2013.

Academia has realized the importance of the demographic transition and is slowly analyzing the characteristics and specific requirements of the so-called *Silver Market*. The following chapters will provide an overview of the SiMa phenomenon and existing attempts to incorporate members of the SiMa into the product development process.

2.2.1 Description of Silver Market

Although it has been identified as an interesting segment, there is no clear and agreed-upon definition of the SiMa. There exist numerous labels used to describe the customer segment of the elderly. The German Wikipedia entry for "Best Agers" lists, besides Silver Agers, the alternative labels "*[...] Generation Gold, Generation 50plus, [...] Golden Ager, Third Ager, Mid-Ager, Master Consumers, Mature Consumers, [and] Senior Citizens*"[53]. This multitude of labels indicates a multitude of approaches adopted to define this market segment. Generally, the segment is defined by a minimum age between 50 and 65 years.[54] The definition of a cut-off value based on chronological age is difficult, because phases of life do not only depend on chronological age but on one's mental and physical state, marital and occupational status, or need for autonomy.[55] Nevertheless, some phases are institutionalized and defined by chronological age. The most drastic is the transition into retirement age, which is highly regulated in most countries (and typically occurs between 60 and 65 years).[56] Typically the minimum age for the definition of the SiMa ranges from 50 to 55 years.[57] For this research, the minimum age is defined as 55 years, which is in line with the definition adopted by most researchers who do not use the entrance into the retirement age as a boundary but rather argue with the changing needs and preferences that typically occur at that age.[58] Additionally, based on a life expectancy of 82 years (which corresponds to the current life expectancy for most industrialized countries), 55 years marks the beginning of the last third of one's life span.[59]

---

[53] Wikipedia contributors 2014.
[54] In a meta-analysis of 67 studies on older consumer behavior by Tongren 1988, the threshold for old age was defined at 49+ (1 study), 55+ (11 studies), 60 / 62+ (9 studies), 64+ (3 studies), and 65+ or older (36 studies). 7 studies did not specify the applied threshold.
[55] Cf. Mayer 1990, pp. 9 & 14; Kohlbacher et al. 2011b, pp. 7ff.
[56] Cf. Kohli 1985, p. 8.
[57] Cf. Szmigin & Carrigan 2001, p. 115; Auken et al. 2006, p. 440; Gassmann & Reepmeyer 2006; Kohlbacher & Herstatt 2008a, p. xi; Fisk et al. 2009, p. 8; Kohlbacher et al. 2011a, p. 193.
[58] Cf. Auken et al. 2006, p. 440; Szmigin & Carrigan 2001, pp. 114f. See also Tongren 1988.
[59] See World Health Organization 2013.

The term *silver service* was first used in Japan in the 1970s. On September 15, 1973, the "Respect for the Elderly Day"[60], the Japanese National Railway introduced silver seats specifically reserved for the elderly.[61] The Japanese word *shirubā* (derived from the English *silver*) refers to the white hair of older people.[62] The name was then applied to other silver products and services and is now a widely-used term.

Compared to previous generations, today's elderly are healthier, more self-reliant, and more demanding of their quality of life.[63] In addition they have the means to afford to become a major driver of economic growth. The median net worth of US households in the age group over 65 is more than double that of the age group 45 to 55.[64] In the UK, the average household expenditures per capita for the age group 65 to 74 are 9 % higher than the average per capita household expenditures. 18.4 % of total household expenditures are made by those of at least 65 years.[65] Individuals over 55 years "[...] are 48 per cent more likely to spend their day shopping, and are 14 per cent more likely to eat out than other adults."[66] German households with residents older than 55 years own 57 % of the net assets although they represent only 44 % of all German households (and 34 % of the population).[67] In Japan, people in their 60s have the highest consumption expenditures, 21 to 39 % above that of the younger non- SiMa age groups.[68]

These highlights show that the SiMa members are, on average an interesting customer segment.[69] Their considerable wealth makes them less price-sensitive. As such, other product characteristics, like quality, convenience, and fostering health are more important as buying criteria.[70] SiMa members are searching for products and services that support them in leading an active and high-quality life.[71] Tempest, Barnatt, and Coupland (2008) suggest a simple segmentation of the SiMa based on the individual's state of health and state of wealth. They show that individuals seek

---

[60] In Japanese called *keirō no hi* and since 1966 a National holiday. It was moved from September 15 to the third Monday in September in 2001 in order to create a long weekend. Cf. Backhaus 2008, pp. 463f..
[61] Cf. Coulmas 2008, p. vi.
[62] Cf. Ogawa 2008, pp. 151f.
[63] Cf. Usui 2008, p. 73.
[64] Cf. United States Census Bureau 2011.
[65] Cf. Office for National Statistics 2012, pp. Table A9.
[66] Szmigin & Carrigan 2001, p. 115.
[67] Cf. Deutsche Bundesbank 2013, pp. Table 1_A_1 & 5_A_1.
[68] Cf. Kohlbacher et al. 2011a, p. 194.
[69] At the same time, poverty among the elderly is a growing problem in developed as well as developing countries. Although it is not to be neglected, it will not be detailed here for reasons of conciseness.
[70] Cf. Arnold & Krancioch 2011, p. 155.
[71] Cf. Usui 2011, p. 334; Reinmöller 2008, p. 160.

products that either improve their state of health, their state of wealth or both. Older customers who are healthy and wealthy seek a high quality of experience.[72] For the purchase, elderlies prefer stores with easy access, sales assistance, and proximity to their home.[73] They are also more likely to be store-loyal and base their purchasing decisions on informal sources of information like recommendations from family and friends.[74]

The majority of companies have not yet targeted the SiMa. In a study among German companies doing business in Japan, Kohlbacher et al. (2011a) found that although more than 90 % acknowledged the medium-term importance of the SiMa, only 45 % saw business opportunities for themselves, and only a small minority is conducting specific marketing (5.4 %) or is developing tailored products (6.5 %).[75]

2.2.2 Product Development for the Silver Market

The difficulty in developing products for the SiMa is that there is a very thin line between a tailored product and one that labels the user as being old. Older people typically perceive themselves as being about 10 years younger, so age stigmatization, whether in product design or marketing, decreases customer satisfaction and will most probably lead to the product being a flop.[76] Levsen (2015) shows that age-based products are often discriminated against in retail markets insofar as they are not provided with shelf access. Of course, non-stigmatization is not possible for all products because some are aids for highly age-specific problems; these include walking frames, adult diapers, and stair lifts. For less age-specific products and services, there exist design criteria to develop ubiquitous products, i.e. they respond to age-specific needs while providing benefits to all age groups, e.g., barrier-free homes, the easy-to-use washing machine Miele Klassik, or cars with an elevated seating position for better circumferential visibility (like the Volkswagen Golf Plus). The most prominent set of design principles is known as *universal design*. Universal design considers the needs and requirements of all potential user groups and does not differentiate between young and old, able and disabled. It aims to integrate all these requirements into one standard instead of creating exceptions for specific user groups.[77] The Center for Universal Design at North Carolina State University defined seven design principles that are generally accepted by product

---

[72] Cf. Tempest et al. 2008, p. 247.
[73] Cf. Arnold & Krancioch 2011, pp. 150ff.
[74] Cf. Moschis 1992b, pp. 245 & 259f.
[75] Cf. Kohlbacher et al. 2011a, pp. 196ff.
[76] Cf. Schmidt-Ruhland & Knigge 2008, p. 107. The effort of non-stigmatizing marketing is visible by the application of terms like *silver agers* or *best-agers*, instead of *seniors* in marketing.
[77] Cf. Gassmann & Reepmeyer 2008, p. 128.

developers: 1) equitable use, 2) flexibility in use, 3) simple and intuitive use, 4) perceptible information, 5) tolerance for error, 6) low physical effort, and 7) size and space for approach and use.[78] Pirkl's (2011) *transgenerational design* follows similar design principles, showing that associations of age and disability are similar. Young people grow old, as able people can become disabled. In the end, both groups need products that enable them to lead a regular life.[79]

Although these guidelines describe how product design should be considered FOR the elderly, they do not define how it can be done WITH them. Suggestions range from asking product designers to envision the mindset of the elderly by simulating typical troubles[80], via the observation of habits and behaviors[81], to active integration in the definition and design process[82]. Research projects like *sentha* ("Everyday Technology for Senior Households", development of products to maintain independent living), *Open ISA* ("Open Innovation Platform for Health-related Services during Old Age"), and *SMILEY* ("Smart Independent Living for the Elderly", technology-based products to assist independent living)[83] have shown that the elderly can efficiently verbalize their specific requirements and that the resulting products could not have been developed by product designers on their own.

Nevertheless, no studies exist that have analyzed whether older people are also creators of age-based innovations and how these innovations can be applied to the creation of silver products.

2.3 Defining Age

At first glance, age seems to be a simple concept. The more time has passed since the birth of a person, the older he or she is. But some people look older than they are and some people do not behave according to their age. Several disciplines of science have developed theories of aging, e.g., biology, psychology, and the social sciences.[84] Age manifests itself in the individual through behavior and the state of the body, but also in conceptions of age in society and culture.[85]

---

[78] Cf. NC State University 1997.
[79] Cf. Pirkl 2011, p. 130.
[80] Cf. Schmidt-Ruhland & Knigge 2008, pp. 114ff.
[81] Cf. Schmidt-Ruhland & Knigge 2008, pp. 109ff.; Helminen 2008.
[82] Cf. Schmidt-Ruhland & Knigge 2008, pp. 111ff.; Östlund 2011, pp. 18ff.
[83] For more information on the projects, please visit the respective project websites: www.sentha.udk-berlin.de, www.tim.rwth-aachen.de/index.php?menu=forschung&inhalt=cpenisa, http://macs2.psychologie.hu-berlin.de/ smiley_projekt/.
[84] See Bengtson et al. 2009a; Backes & Clemens 2008, p. 92.
[85] See Staudinger & Häfner 2008.

2.3.1 Shortcomings of Chronological Age and Alternative Age Measurements

A person's chronological age is the time in years that has elapsed since his or her birth. This measure is applied in almost all cultures except for some Asian ones which measure chronological age from conception.[86] The basic stages in life, like childhood, education, work life, and retirement, are typically defined according to chronological age.[87] According to the life course principle, aging occurs at any time from birth until death, and it is defined through biological, psychological, and social processes.[88] Although chronological age is a good indicator of the general characteristics of a specific age for a larger population, it does not reliably describe someone's individual capabilities and preferences.[89] Some people are vital and in the best of health at 90 years, while others are in delicate health in their 50s. Some older people experience a dramatic loss of cognitive capabilities while others perform as well as much younger people. A good age measure must be "[...] more sensitive to individual differences."[90]

In fact, although the underlying reasons behind aging have been intensively researched, they are not yet completely understood. Gerontology, which is the science of the biological, psychological, and social aspects of aging, has been labeled "*data-rich but theory-poor*"[91]. A detailed overview of existing theories on the reasons for aging cannot be provided, because the required depth would be beyond the scope of this work.[92]

Several alternative age measures have been suggested, mainly biological age and functional age. Biological age focuses on the health status of an individual and assesses relative age based on the presence of specific biomarkers.[93] The assessment of biological age requires profound medical knowledge, time, and direct contact with the subject under investigation, which makes its application in a business environment almost impossible.

In addition to the health status, functional age also takes cognitive capacities and behavior into account. Studies on functional age typically include anthropometric,

---

[86] Cf. Charness & Krampe 2008, p. 244.
[87] Cf. Kohli 1985, p. 2; Mayer 1990, p. 14.
[88] Cf. Bengtson & Allen 1993, pp. 470ff.
[89] Cf. Super 1994, p. 254; Sudbury & Simcock 2009, p. 23.
[90] Settersten, Jr. & Mayer 1997, p. 239.
[91] Bengtson et al. 2009b, p. xxi.
[92] The interested reader is referred to comprehensive standard works, like Bengtson et al. 2009a; Hofer & Alwin 2008; Hooyman & Kiyak 2011; Schaie & Willis 2011 or gerontology journals, especially Age (ISSN: 0161-9152), Age and Ageing (ISSN: 0002-0729), and The Journals of Gerontology (ISSN 1079-5006 and 1079-5014).
[93] Cf. Ludwig & Smoke 1980; Baker, III. & Sprott 1988, p. 228.

dental, sensorimotor, physiological, cognitive, psychosocial, and behavioral variables.[94] Since there is no generally accepted definition of functional age, the selection of biomarkers varies widely, based on availability and functional outcome. In a review of empirical studies on measuring functional age, Anstey, Lord, and Smith (1996) analyzed 24 studies using 177 different biomarkers.[95] The effort required for the assessment of functional age again makes its implementation in a business context unprofitable.[96]

All age measurements are oriented on a standardized progress of age through a comparison with the average. Therefore, they are all linked to chronological age and use it to make relative statements ("You have the biological age of a 50-year-old man.").

2.3.2 Cognitive Age

A measurement that is more reliable than chronological age (with regards to capabilities and preferences) but is easier to estimate than functional age was required. Kastenbaum et al. (1972) realized that people often perceived their own age differently than their true chronological age. They introduced the "ages-of-me" model which took into account self-evaluations of several dimensions. Building upon this model, Barak and Schiffman (1981) suggested the use of a person's self-perceived cognitive age, based on the evaluation of his/her feel-, look-, do-, and interest-age. The age-dimensions relate to "*emotional (feel-age), biological (look-age), societal (do-age), and intellectual (interest-age)*"[97] aspects of the individual. Following studies have shown that cognitive age is superior to chronological age in explaining the self-perceptions and behaviors of older consumers.[98] Cognitive age has been associated with self-respect and reputation[99], need for security[100], internal locus of control[101], fashion interest[102], willingness to try new brands[103] and interest in seeking information[104]. More recent studies have also shown the usefulness of

---

[94] Cf. Anstey et al. 1996, pp. 252ff.
[95] Cf. Anstey et al. 1996, pp. 250ff.
[96] Cf. Kohli 1985, p. 14.
[97] Barak 2009, p. 3.
[98] Cf. Kohlbacher & Chéron 2011, p. 180; Wilkes 1992, p. 292.
[99] Cf. Cleaver & Muller 2002, pp. 238f.; Wilkes 1992, p. 297; Sudbury & Simcock 2009, p. 31.
[100] Cf. Sudbury & Simcock 2009, p. 30.
[101] Cf. Hubley & Hultsch 1994, p. 433.
[102] Cf. Wilkes 1992, p. 297.
[103] Cf. Stephens 1991, p. 44.
[104] Cf. Gwinner & Stephens 2001, p. 1046.

cognitive age for segmentation.[105] Since its correlation with the most prominent demographics (like gender, marital status, race, education, and income) is low, it provides valuable information not captured by those demographics.[106]

The age difference between cognitive age and chronological age for older consumers is typically between 8 and 15 years.[107] Although it has been predominantly researched in Western, Anglophone countries, research could also establish functional, conceptual, and measurement equivalency for Eastern (e.g., China, Korea, Japan) and non-Anglophone countries (e.g., Brazil, France, Croatia).[108] These results suggest that cognitive age is truly *"the global age-identity construct"*[109].

2.4 Effects of Aging

Although the reasons for why organisms age are not fully understood yet, the effects of aging on the human body have been investigated. Aging affects the physical, sensory, and cognitive capabilities. As such, developers of age-based products must take these into account. Although the following effects can be regarded as generally applicable, the timing of occurrence and intensity of the effects can differ greatly between individuals.[110]

The human body changes with age, leading to anatomical changes, like an increase in hand thickness, the width of thumbs and the index fingers, and a reduced flexibility of the cervical spine and wrists.[111] Minute motor activity is reduced, along with the grip strength and the length a firm grip can be maintained.[112] The decrease in muscular mass leads to a decline in overall physical strength.[113] Lung volume and pulmonary elasticity are reduced, and the rate of cerebrovascular and cardiovascular diseases steadily increases.[114]

Sensory capabilities are also affected. Several aspects of hearing (e.g., ability to hear high-pitched sounds, tolerance for background noises), vision (e.g., light

---

[105] Cf. Sudbury & Simcock 2009, p. 32; Auken & Barry 2009, pp. 323f.
[106] Cf. Henderson et al. 1995, p. 455.
[107] Cf. Cleaver & Muller 2002, p. 238; Hubley & Hultsch 1994, p. 416.
[108] See Auken et al. 2006; Barak 2009; Barak et al. 2011.
[109] Barak 2009, p. 5.
[110] Cf. Backes & Clemens 2008, p. 93.
[111] Cf. Bleyer et al. 2009, p. 11.
[112] Haigh 1993, pp. 9ff.
[113] Cf. Moschis 1992b, p. 96.
[114] Cf. World Health Organization & US National Institute of Aging 2011, pp. 18f.

requirements, visual acuity, color perception), and tactile sensation (e.g., number of tactile corpuscles, skin sensibility) are negatively affected by aging.[115]

Some cognitive capabilities seem to be immune to aging. Studies have shown that crystallized intelligence (e.g., general knowledge, vocabulary) does not differ among age groups, in contrast to fluid intelligence (e.g., short-term memory, problem-solving).[116] The information processing speed and capacity is reduced, and the elderly require more time to fulfill complex tasks.[117] The ability to quickly switch between tasks is reduced and the time required to learn new schemata is increased.[118] On the other hand, reasoning about social conflicts (Grossmann et al. (2010) call it wisdom) actually improves with age, meaning that older people perform better in mediation.[119] The decline of cognitive capabilities is not a phenomenon of old age. Rather, it begins when adults are in their 20s and 30s.[120]

Physical, sensory, and cognitive decline result in a higher susceptibility to accidents and diseases among elderlies. But they have strategies to cope with some deficits. In a working environment, older employees tend to solve stressful tasks more slowly but with a greater precision.[121]

Research on older consumers has shown that the elderly clearly have a different set of attitudes and values than younger consumers. They are less selfish and show compassion for others. The importance of this attitude becomes even more important because it is already higher among the currently younger age cohorts.[122] Safety and security are two very important values for the elderly, as well as a sense of purpose, social connectedness, and independence/need for autonomy.[123] Their increased desire for security and safety does not necessarily make older people completely risk averse. In an investigation of the influence of perceived risk on high-involvement purchasing decisions, only physical risk was higher among the elderly. All other risk types, i.e., functional, financial, social, psychological, and time risk, did not show significant differences.[124]

Regarding the marketing of products, Wolfe (1994) identified five underlying key values that drive product selection among older consumers: 1) autonomy and self-

---

[115] Cf. Saup 1993; Gruca & Schewe 1992, pp. 19f.; Fisk et al. 2009, pp. 15ff.
[116] Cf. Horn & Cattell 1967, p. 107; Sorce 1995, pp. 470ff.; Fisk et al. 2009, p. 242.
[117] Cf. Grossmann et al. 2010, p. 7247; Sorce 1995, p. 467.
[118] Cf. Schapkin 2012, p. 82.
[119] Cf. Grossmann et al. 2010, p. 7249.
[120] Cf. Salthouse 2009, p. 507.
[121] Cf. Schapkin 2012, p. 82.
[122] Cf. Plutzer & Berkman 2005, p. 80.
[123] Cf. Dychtwald & Flower 1990; Schewe 1991, pp. 61ff.; Kohlbacher et al. [in press].
[124] Cf. Simcock et al. 2006, pp. 357ff. & 365.

sufficiency, 2) social and spiritual connectedness, 3) altruism, 4) personal growth, and 5) revitalization.[125] Marketers should emphasize comfort, convenience, and a good experience when targeting older consumers.[126] The importance of different information sources for making purchasing decisions remains unclear in the literature. While some authors show that older consumers rely more on informal sources (e.g., family, friends, and neighbors)[127], others argue that formal sources (e.g., sales assistants, mass media) are more important.[128] Wolfe (1994), on the other hand, suggests that older consumers rely mostly on their own subjective experience, rather than on external sources.[129] Schiffman and Sherman (1991) confirm this suggestion in their description of the new-age elderly.[130]

2.5 Age and Innovative Behavior

In consumer research, innovativeness or innovative behavior is defined as the early adoption of new products, and not as the actual development of new or improved products.[131] Under this adoption-oriented view, age has a negative impact on consumer innovativeness,[132] although some studies have failed to confirm a significant relationship.[133] Cognitive age also plays a relevant role, because older people who perceive themselves as younger are typically more likely to adopt new products and try new brands.[134]

Innovative behavior by users in terms of the development of new products in conjunction with age has not yet been the focus of research. The existence of the phenomenon across all age groups has been indicated by Hippel, Jong, and Flowers (2012) and Ogawa and Pongtanalert (2011) (see also chapter 3.2 below). Most of the insights into the relationship between age and innovative behavior stem from literature on organization and human resources, which focuses on the capabilities of employees in R&D departments. Inventive output of R&D personnel over age shows an inverted u-shape with a climax reached in the early 30s and a significant drop

---

[125] Cf. Wolfe 1994, p. 32.
[126] Cf. Wolfe 1994, pp. 35f.; Schiffman & Sherman 1991, pp. 189f..
[127] Cf. Lumpkin et al. 1989, p. 182.
[128] Cf. Arnold & Krancioch 2011, pp. 150ff.; Tongren 1988, p. 148.
[129] Cf. Wolfe 1994, p. 35.
[130] Cf. Schiffman & Sherman 1991, p. 192.
[131] Cf. Im et al. 2003, p. 61; Rogers 2003, p. 247; Roehrich 2004, p. 671; Midgley & Dowling 1978, p. 229.
[132] Cf. Im et al. 2003, p. 69; Steenkamp et al. 1999, p. 65.
[133] Cf. Schreier & Prügl 2008, p. 343.
[134] Cf. Stephens 1991, p. 44.

after the age of 40.¹³⁵ Eisfeldt (2009) claims that the chance to be an innovator decreases 3 % per year for individuals with a high education. Therefore, a 40-year-old is 26 % less likely to become an innovator than a 30-year-old.¹³⁶ The main driver seems to be a decrease in creativity, but the advantage of greater experience can compensate for most of this decrease.¹³⁷ Bergmann, Prescher, and Eisfeldt (2006) even found no significant relationship between age and inventive output among engineers in SMEs.¹³⁸ One factor that is usually not included in these studies was the fact that engineers potentially switch into roles along their career path in which their focus is no longer on product development but rather on managing a department.¹³⁹ A look at the output of academic scholars, who usually are not affected by such changing job requirements, nevertheless shows a very similar pattern. Their output is also an inverted u-shape with a climax depending on the specific discipline. Some disciplines are characterized by a peak at the late 20s or early 30s, e.g., mathematics and theoretical physics, while at others the peak is not reached before the late 40s, e.g., history, philosophy, and medicine.¹⁴⁰

An analysis of Thomas Edison's patents provides an excellent example of the inventive output of an individual. Over the course of his life, Edison filed 1,093 patents from the age of 21 until his late 80s. Many of his inventions heavily influenced people's lives, including the first commercially practical light bulb, the phonograph, the motion picture camera, and the stock ticker.¹⁴¹ The graph of his patents (see Figure 5 below) shows a sharp increase in his early 30s and a peak at the age of 35, when he filed 106 patents under his name. Besides a gap between the age of 45 and 50, the numbers then slowly decrease but remain generally stable.

---

[135] Cf. Hoisl 2007, p. 21; Oberg 1960, pp. 251ff.
[136] Eisfeldt 2009, p. 166. $(1 - 0.03)^{10} = 0.74$.
[137] Cf. Oberg 1960, p. 253; Adenauer 2002, p. 42.
[138] Cf. Bergmann et al. 2006, p. 25. Oberg 1960 found a similar result when he separated R&D employees and engineers in his sample. While R&D employees showed the expected peak in the mid-30s, the evaluation of the output of the engineers steadily increased with age, with the age groups 51-55 and 56-60 showing the highest values. Cf. Oberg 1960, pp. 253ff.
[139] Cf. Bergmann et al. 2006, p. 19.
[140] Cf. Simonton 1988, pp. 252 & 262.
[141] Cf. Wikipedia contributors 2013a.

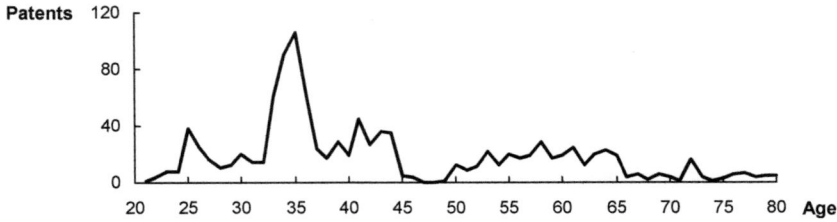

Figure 5: Number of Thomas Edison's US Patents by Age, based on Execution Date[142]

This example shows that although inventive output is typically highest before the age of 40, people are still able to develop meaningful innovations throughout their life.

2.6 Interim Conclusions

Chapter 2.1 has shown that the demographic shift impacts all countries across the globe. Western countries have already experienced a sharp increase in the share of older people as the baby boomer generation approaches retirement age. But this demographic shift should not be regarded solely as a threat to social systems. Today's elderly are healthier, better educated, and more independent than any generation before them. They demand tailored products without the stigmatization of being old. This SiMa is an attractive market for companies, but its approach remains challenging.

The underlying reasons for aging are not yet fully understood but the key effects of aging have been identified. There exist multiple explanations regarding how the physical, sensory, and cognitive changes affect the preferences and behavior of the elderly. Researchers and practitioners are now slowly acknowledging the fact that there does not exist a *typical senior*. Rather, the SiMa is more heterogeneous than younger customers. This is manifested, for example, in differences in the self-perceived cognitive age.[143]

In research studies on user innovation, data on age is rarely provided, but representatively large-N studies have shown that the phenomenon exists across all age groups.[144] The relationship of age and inventive output has only been analyzed

---

[142] Own illustration. N = 1,093. Source of data: http://edison.rutgers.edu/patents.htm, accessed on June 26, 2013. Execution date is the date on which the inventor signs the application for filing at the US Patent Office.
[143] Cf. Moschis 1992a, p. 18; Backes & Clemens 2008, p. 343; Arnold & Krancioch 2011, p. 149; Sudbury & Simcock 2011, p. 196.
[144] Cf. Hippel et al. 2012; Ogawa & Pongtanalert 2011.

in the labor sciences. It has been shown, that creativity decreases past the age of 40, but this loss is often compensated with greater experience and social capabilities. These studies are, by design, limited to the boundaries of the firm and individuals above the retirement age are not included. Insights on the innovative behavior of people past the age of 60 are therefore almost not available.

3 Users as Main Source of Innovations

This chapter will describe the important role of users in the innovation process and how this discovery led to the lead user theory. The characteristics of user innovators and their innovations will be described. Influencing factors of innovative behavior and the scarce research focusing on the relationship between innovative behavior and age will also be identified.

3.1 Development of User Innovation Research

Traditionally, manufacturers have been seen as the source of innovation and new product developments. Since Schumpeter's work, the driving force behind product innovations has been assigned to producers and policy makers.[145] In Schumpeter's view, consumers had the mere role of selecting among the competing offers. Later, he focused even more strongly on the role of producers, who needed to constantly improve and redefine products, processes, and organizations in order to stay ahead of the competition. Schumpeter later labeled this permanent firm-driven improvement process *Creative Destruction* and identified it as the driving force of capitalism.[146]

That consumers might play a more important role was hypothesized during research on the diffusion of products. Rogers (1962) introduced an idealized diffusion curve for products that distinguished adopter types according to their innovativeness. He defined innovativeness as "*[...] the degree to which an individual is relatively earlier in adopting new ideas than the other members of his system.*"[147] Rogers still assumed that manufacturers are responsible for new product developments and that consumers are innovative merely if they adopt these new products early. According to his diffusion curve, Rogers derived five adopter categories: innovators (2.5 %), early adopters (13.5 %), early majority (34 %), late majority (34 %), and laggards (16 %).[148] In the context of product diffusion, innovators are especially important,

---

[145] See Schumpeter 1934.
[146] Cf. Schumpeter 1942, pp. 82ff.
[147] Rogers & Shoemaker 1971, p. 27.
"Relatively earlier" was based on the actual tie of adoption, not on the perceived relative adoption time.
[148] Adopter categories were defined based on their standard deviation from the average adoption time. Innovators deviated negatively at least two standard deviations from the mean, early adopters between one and two negative standard deviations, and laggards at least one positive standard deviation. Adopters of the early majority and late majority are all within one standard deviation from the mean differ according to whether they deviate positively or negatively from the mean.

because they adopt products without requiring confirming positive reviews from other users, and they are the first users who can provide feedback.[149] Comparing early and late adopters, it was found that early adopters differ in socioeconomic status (e.g., higher education, higher social status), personality variables (e.g., greater rationality and intelligence, higher motivation), and communication behavior (e.g., higher degree of opinion leadership).[150] Although Roger's diffusion curve has been criticized[151], it nevertheless is widely accepted and is still in use even 50 years after its first introduction. It is especially relevant in the field of marketing.[152] Since Roger's innovators do not fulfill the definition of innovation in chapter 1.4, they are regarded as "lead adopters" rather than true innovators. Nevertheless, his findings on the diverse needs of adopters laid the foundation for research on user innovations that would emerge in the following decade.

3.2 Users as the Main Functional Source of Innovation

Although Adam Smith already identified the existence of user innovators by pointing out that "*a great part of the machines made use of in those manufactures in which labour is most subdivided, were originally the inventions of common workmen, who, being each of them employed in some very simple operation, naturally turned their thoughts towards finding out easier and readier methods of performing it*"[153], it took almost 200 years for researchers to begin to systematically study the phenomenon of user innovation.

The first quantitative studies that described the phenomenon of user innovations (which occurred more by accident, because user innovations were not the focus of these studies) were published in the 1960s. Enos (1962) analyzed the development of cracking processes in petroleum refineries between 1913 and 1957. He discovered that most of these innovations were actually introduced by user firms and were later adopted from manufacturers of the equipment. Freeman (1968) found similar results for processes in the chemical industry, where 70 % of the improvements were introduced by user firms.[154] The first two articles with a focus on

---

[149] Cf. Midgley 1977, p. 49.
[150] Cf. Rogers 2003, pp. 269f.
[151] E.g., for its assumption of a normal distribution, the hard and seemingly arbitrary separation of categories, and the measurement of innovativeness via the time of adoption of just one specific product.
[152] Cf. Midgley 1977, pp. 53f.
[153] Smith 1778, p. 12.
[154] Cf. Freeman 1968, p. 44.

the importance of users in the development process were published in 1976 and were both written by Eric von Hippel, kick-starting a new research field.[155]

Hippel analyzed innovations in the semiconductor industry and in the field of scientific instruments and found that the majority of innovations were developed by users. Based on this observation, Hippel demonstrated that the *"functional source of innovation"*[156] varies between industries and product categories and can reside within users, manufacturers, suppliers, and others.

The source of innovation can be predicted by analyzing the distribution of the expected benefits. Innovations are most likely created by those players who expect the highest benefit from the innovation.[157] As was stated in chapter 1.4 above, users benefit from using an innovation while manufacturers benefit from selling (or licensing) it. The net benefit of an innovation is influenced by the heterogeneity of needs, effectiveness of patents, and stickiness of information.[158]

Heterogeneity of Needs

Heterogeneity of needs can be considered high if the customer requirements differ strongly between segments (or even individual customers). A manufacturer must calculate potential revenues and cost before it can make the decision to invest in a new product. The revenues depend heavily on the number of customers, but customer needs can be very different in some product categories. When customer requirements differ strongly, the product must be tailored to specific segments to be attractive, but this decreases the size of the potential customer base. Manufacturers usually decide to design for a larger customer base and not to cater to specific demands, which increases the benefits for the manufacturer but decreases the individual benefits for the user.[159] In contrast, user innovators do not need to worry about market demand or heterogeneity of needs. If a user has a specific need and there is no product or service in the market to fulfill it, the user just has to answer two very simple questions: 1) Does my benefit from using the innovation outweigh my cost of developing it (including time, material cost, tools, etc.)? and 2) Am I able to realize the innovation? If the user can answer both questions positively, it is beneficial for him to innovate. In most cases, free revealing of information regarding

---

[155] See Hippel 1976a, 1976b.
[156] Hippel 1988, p. 3. For a detailed description of the different functional sources of innovation, the reader might especially focus on chapter 3 of Hippel 1988.
[157] Cf. Hippel 1988, pp. 5f.
[158] Cf. Tinz 2007, p. 104.
[159] Cf. Hippel 2005a, p. 51; Tinz 2007, p. 89.

the innovation – so that other users can adopt, and potentially improve, the innovation – will most probably increase the user's benefit even more.[130]

Effectiveness of Patents

A patent grants the patent holder the right to exclude others from exploiting (through manufacturing, usage, sale, or import) the patented invention for a certain time frame.[161] The grant of a patent is tied to costs for the application and enforcement of the patent. Studies have shown that even for manufacturers, patents are not always useful to capture royalties or exclude imitators because of the high patenting cost involved and because of the (in some cases) weak patent protection system.[162] Also, in most industries (except chemicals and pharmaceuticals), patenting does not affect the innovative output of firms.[163] On the contrary, industries, like the software and electronics industry, have witnessed the advent of *"[...] patent thicket[s]: an overlapping set of patent rights requiring that those seeking to commercialize new technology obtain licenses from multiple patentees"*[164] in recent years. Manufacturers, therefore, often also use alternative approaches to protect and monetize their innovations. A common approach is to use trade secrets. Trade secrets can be kept secret even after the commercialization of the product because they are either embodied in the product itself and cannot be reverse-engineered without extensive effort (e.g., the recipe for Coca Cola) or the innovation is inherent in the production process and its equipment, which can be protected through the factory walls.[165] Hippel found that firms that were able to protect their process equipment innovations anticipated higher benefits from innovations and were more likely to innovate.[166] This is usually easier for users than for manufacturers because the latter must also reveal process information to support potential adopters of the innovation.

As mentioned, patents are often not profitable for users, but even trade secrets might not be feasible because other users usually have comparable know-how and might be willing to reveal it.[167] Free revealing of proprietary knowledge by users has been observed in many studies[168] but is especially relevant for *open source* projects. Free revealing entails *"[...] that all existing and potential intellectual property rights to that*

---

[160] Cf. Harhoff et al. 2003; Hippel 2005a, pp. 77ff.; Raasch et al. 2008, pp. 383ff.
[161] Cf. World Intellectual Property Organization, p. 17.
[162] Cf. Wilson 1975; Taylor & Silberston 1973.
[163] Cf. Mansfield 1986, p. 180.
[164] Shapiro 2001, p. 119.
[165] Cf. Hippel 1988, p. 54.
[166] Cf. Hippel 1988, p. 5.
[167] Cf. Hippel 2005a, p. 10.
[168] See Urban & Hippel 1988; Ogawa 1998; Morrison et al. 2000; Lilien et al. 2002; Lüthje 2004.

information are voluntarily given up by that innovator and all interested parties are given access to it—the information becomes a public good."[169] This strategy can be the most profitable for users under certain conditions: Innovators can already gain personal benefit from the process of innovation itself (e.g., learning, personal joy, reputation), other users might further improve the innovation and share the results, and the free revealing of information can assist the diffusion, resulting in additional benefits through network effects for the innovator.[170] Additionally, a user whose idea is adopted by many others can thereby define a standard, which is based on a tailored solution for his or her specific preferences and can become a permanent advantage.[171] This can motivate users to be the first to reveal innovations.[172] Based on the observation, that under some conditions, users invest privately into innovations and then freely reveal them turning them into public goods, Hippel and Krogh (2003) coined the term *private-collective* innovation model.[173]

In summation, if benefits from patents and trade secrets are expected to be high, manufacturers are likely to innovate, because they can monetize their proprietary knowledge and gain a competitive advantage. If patent costs are high and protection is low, users are more likely to innovate because they can draw benefits from the innovation process itself, and it might even be profitable for them to freely reveal their innovation.

Stickiness of Information

The results of a product development process can only be as good as the information on needs and technology that was available for initial input. It follows the principle of *garbage in, garbage out*. Innovators require valuable and relevant information to develop a successful new product, process, or service. Users best understand their needs and the contextual factors of a product's use. Manufacturers usually best understand solution knowledge and technological aspects of a product.[174] If the transfer of information requires no transaction cost, any player could innovate. But transaction costs exist and often reach a prohibitive level, making it too costly to make an innovation profitable. If information is difficult to transfer from one player to another, the information is *sticky*. Hippel (1994) defined "*[...] the stickiness of a given unit of information in a given instance as the incremental expenditure required to transfer that unit of information to a specified locus in a form usable by a given*

---

[169] Harhoff et al. 2003, p. 1753.
[170] Cf. Harhoff et al. 2003, p. 1757.
[171] Cf. Allen 1983, pp. 17ff.
[172] Cf. Harhoff et al. 2003, p. 1757.
[173] See Hippel & Krogh 2003.
[174] Cf. Hippel 2005a, pp. 66f.

information seeker."[175] The reasons why information can be sticky are manifold. One of the most often mentioned reasons is that individuals hold certain knowledge which is implicit and therefore cannot be transferred. A popular example is riding a bike: A person who can successfully ride a bike is not necessarily able to explicitly explain how he or she is considering speed, balance, inclination, and steering angle. According to Polanyi (1958) many human skills and experiences are of such implicit nature; he established the term *tacit knowing* for them.[176] Another reason why information can be sticky is that, in some cases, the recipient requires specific prior related knowledge in order to understand the information and put it into context. The extent of prior related knowledge strongly relates to absorptive capacity, which determines innovative capabilities.[177] Finally, even if information is explicit and the recipient has the absorptive capacity to assimilate it, it is possible that the amount of required information is so huge that important points get lost during the transfer. In this case, although the successful transfer of single pieces of information is not costly, the total amount is.[178]

We can see that the stickiness of information strongly influences the innovation cost. Therefore, it heavily influences the expected net benefit of an innovation. Sticky information can exist on the manufacturer's as well as the user's side. If sticky information is present, the innovation will probably be developed where it is present.[179]

As described, heterogeneity of needs, effectiveness of patents (and with it the attractiveness of freely revealing information), and stickiness of information determine where the functional source of innovation most probably will be. Several studies have also found that the importance of users increases the more fundamental the type of innovation is. First-of-type innovations especially are almost exclusively developed by users.[180] Manufacturers then step in later and develop "*[...] functional substitutes for existing user innovations.*"[181]

---

[175] Hippel 1994, p. 430.
[176] Cf. Polanyi 1958, pp. 48f.
[177] See Cohen & Levinthal 1990.
[178] Cf. Hippel 1994, pp. 68f.
[179] Cf. Hippel 1994, p. 430.
[180] Cf. Hippel 1976b, p. 222; Shaw 1985, p. 290; Riggs & Hippel 1994, p. 466; Shah 2000, p. 9.
[181] Slaughter 1993, p. 86.

Table 1: Selected Studies on User Innovations

Study	Product	Sample Characteristics	Study Object	User Innovation
B2B				
Freeman 1968	Chemical process equipment	720 innovations	Innovation	70.0 %
Hippel 1976b	Scientific Instruments	111 innovations	Innovation	81.0 %
Shaw 1985	Medical equipment	34 innovations from 11 firms	Innovation	76.0 %
Voss 1985	Software applications	63 users and suppliers of software	Innovation	32.0 %[a]
Vanderwerf 1990	Wire preparation equipment	20 innovations of construction equipment	Innovation	16.7 %
Riggs & Hippel 1994	Scientific instruments	64 innovations related to Auger and Esca	Innovation	44.0 %
Urban & Hippel 1988	Software applications	136 users of PC-CAD software	User	23.0 %
Herstatt & Hippel 1992	Pipe hangers	74 employees of pipe hanger installing companies	User	36.0 %
Morrison et al. 2000	Library information systems	122 libraries using OPAC systems	User	26.0 %
Franke & Hippel 2003	Software application	131 administrators of Apache server software	User	19.1 %
Jong & Hippel 2009	Process equipment and software	498 Dutch high-tech SMEs	User	54.0 %
B2C				
Shah 2000	Board sporting equipment	57 innovations in skateboarding, snowboarding, and windsurfing	Innovation	58.0 %[b]
Franke & Shah 2003	Extreme sporting equipment	197 members of extreme sports clubs	User	32.1 %
Franke et al. 2006	Kite surfing	456 users of kite surfing equipment	User	31.7 %
Lüthje et al. 2002	Mountain bikes	291 members of mountain bike clubs	User	38.7 %[c]
Lüthje 2004	Outdoor equipment	153 customers of mail order company for outdoor equipment	User	37.3 %[d]
Tietz et al. 2005	Kite surfing	157 users of kite-surfing equipment	User	41.0 %[e]

a) In 32 % of the cases, users provided the idea; in 20 % of the cases users developed a working product.
b) For first-of-type innovations, share of user innovations was at 100 %.
c) 38.7 % of users had an idea, 19.2 % developed a prototype.
d) 37.3 % of users had an idea, 9.8 % developed a prototype.
e) 41.0 % of users had an idea, 26.0 % developed a prototype.

In some cases it is possible that the tasks of the innovation process are separated, and each step is then executed by the player with the relevant information and capabilities needed to execute it at the lowest cost. This iterative process only works if the cost of coordinating and transferring knowledge between the players is lower than the cost of transferring all required information to one player.[182] Usually tasks focusing on needs will be conducted by users, while tasks focusing on solutions tend to be conducted by manufacturers during cooperative product-development

---

[182] Cf. Hippel 1994, p. 433.

processes.[183] A very recent finding is that sometimes users of a product are also employees of the manufacturer of the product and can therefore serve both roles and bridge the gap between both sides.[184]

After Hippel discovered the importance of users in the product development process, other researchers turned their attention to this phenomenon. Early studies focused on the existence and importance of user innovations for industrial goods, especially in scientific instruments[185], medical equipment[186], software applications[187], and construction equipment[188]. At the start of the new millennium, researchers found that the phenomenon of user innovators also exists for consumer products. Research on consumer products focused almost exclusively on newly developing sports and outdoor activities[189]; the few exceptions were not published in any major journal[190]. Across all studies, the share of innovation developed by users was usually well above 30 % and was highest for industrial products where the share of user innovation was up to 81 % (see Table 1 above). But besides the high user share among innovations, many users also innovate. Table 1 above shows several studies where users were the main study object and the share of users that innovated ranged from 19 % to 54 %. These high shares resulted from the selection of products where a high share of innovations could be expected. Nevertheless, user innovators are not limited to specific niches and may also exist in ordinary everyday life. Three nationwide, representative studies in the US, UK, and Japan have found that user innovators exist in all three countries, although to a different degree. The highest share was found in the UK (N = 1,173) with 6.1 %, followed by the US (N = 1,992) with 5.2 %, and Japan (N = 2,000) with 3.7 %.[191] There does not exist an explanation

---

[183] Cf. Hippel 2005a, p. 72.
[184] See Schweisfurth & Raasch 2012 and Schweisfurth 2013.
[185] See Hippel 1976b; Riggs & Hippel 1994.
[186] See Shaw 1985; Lüthje 2003.
[187] See Voss 1985; Urban & Hippel 1988; Morrison et al. 2000; Franke & Hippel 2003.
[188] See Vanderwerf 1990; Herstatt & Hippel 1992.
[189] Shah 2000 looked into board sports, Franke et al. 2006 and Tietz et al. 2005 focused on kite surfing, Lüthje et al. 2002 on mountain bikes, Baldwin et al. 2006 on rodeo kayaking, Lüthje 2004 on general outdoor equipment, and Franke & Shah 2003 on extreme sporting equipment in general.
[190] See for example the following: Marchi et al. 2011 analyzed whether user innovators existed in an online brand community of the motorcycle brand Ducati. Füller et al. 2007 describes different user innovator types in an online community for a physical product in a mature market: basketball shoes. A research project at the Technical University of Munich analyzed the possibility to identify lead users in virtual communities for food and beverage manufacturers (see Casper & Reichert 2008; Jiptner et al. 2009) and for the elderly (see Baumbach & Schmidle 2008).
[191] Cf. Hippel et al. 2012, p. 1675 for UK figures and Ogawa & Pongtanalert 2011, p. 6 for the US and Japan. In an earlier study on general user innovation among UK consumers aged at least 15 years, the share of user innovators was even higher at 8.0 % (N = 2,109) (cf. Flowers et al. 2010, p. 16).

yet for why the shares are different among countries. Similar research also needs to be conducted in lower-income countries to compare results.[192] But one can already state that, if around 5 % of people in a country invest their time, money, and resources, their combined expenditures on product development are considerable. Hippel, Ogawa, and Jong (2011) estimated that the annual expenditures by consumer innovators compared to the spending of commercial enterprises on consumer products is 144 % in the UK, 33 % in the US, and 13 % in Japan.[193]

3.3 Characteristics of User Innovators

The reasons why users innovate and the typical characteristics of user innovators are manifold and will be the focus of this chapter.

While Rogers (1962) based his definition of innovativeness on the adoption behavior of users, innovativeness is usually considered to be a personality trait, which means that it is a stable disposition of an individual that distinguishes it from others.[194] Midgley and Dowling (1978) defined innovativeness as "*[...] the degree to which an individual makes innovation decisions independently of the communicated experience of others.*"[195] The problem with this *innate innovativeness* is that its measurement is still based on adoption behavior and not necessarily on the development of new products. Also, it assumes that a person holds the same degree of innovativeness regardless of context. Hence, Goldsmith and Hofacker (1991) proposed a domain-specific innovativeness that allows for a person to be very innovative in a certain product field where he has a lot of experience, interest, or the like and might be not innovative at all in another.[196] The assumption that motivation, qualification, and innovative behavior must be interpreted within the specific context of a product field has been generally applied by researchers and has been shown to correlate best with innovative behavior and only weakly with general personality traits.[197]

Although most studies focus on the relationship between personal characteristics and new product adoption behavior, there are a few demographic qualities that have

---

[192] Cf. Hippel et al. 2011, p. 28.
[193] Cf. Hippel et al. 2011, p. 30.
[194] Cf. Morrison 1996, p. 8; Midgley & Dowling 1978, p. 229.
[195] Midgley & Dowling 1978, p. 235.
[196] Cf. Goldsmith & Hofacker 1991, p. 219.
[197] Cf. Lüthje 2004, p. 685; Im et al. 2003, p. 63; Bearden et al. 2011, pp. 109ff.; Roehrich 2004, p. 675; Hoffmann & Soyez 2010, p. 780; Venkatraman 1991, pp. 62f.

been found to be linked to innovativeness.[198] Innovative users are typically younger, more highly educated, and more technically trained than average citizens. They are also more likely to be male and single.[199] Some other studies did not find a clear link between personal characteristics and user innovativeness. As such, it remains difficult to identify potential innovators by demographic data alone.[200]

Studies that have incorporated motivational factors have found that those factors can help to explain the innovative behavior of respondents. Financial rewards or other extrinsic motivators are typically not relevant for user innovators and might even have a negative effect on innovative or cooperative behavior.[201] Except in rare cases, when user innovators want to commercialize their innovations and profit directly from them, they innovate because they want to improve their performance within a specific activity.[202] In addition to the improvement of one's own performance, joy in the innovation process itself, helping others, and reputation effects foster innovative behavior.[203] A high level of intrinsic motivation may even outweigh a lack of technical expertise if the user innovator invests sufficient resources in trial-and-error.[204]

Besides these general characteristics, user innovators often also require highly context-specific competencies or know-how (e.g., openness to new technologies, access to special technologies, and use experience) to be able to successfully carry out the development of an innovation.[205] These characteristics are so specific by nature that it is almost impossible to define them on a general level.

3.4 Lead User Theory

As noted in the previous chapters, user innovators exist across many product categories. In 1986, Hippel published an article stating, that there exists an even smaller group of users who are distinct from other users and even user innovators. These users are so advanced in the execution of certain activities that they are far

---

[198] An overview on studies regarding the relationship between personal characteristics and new product adoption behavior can be found in Im et al. 2003, p. 64.
[199] Cf. Hippel et al. 2011, p. 28; Steenkamp et al. 1999, p. 63; Eisfeldt 2009, pp. 150ff.; Midgley & Dowling 1993, p. 619.
[200] Cf. Steenkamp et al. 1999, p. 63 found no relationship regarding education or income, Im et al. 2003, pp. 67ff. found no relationship regarding income, age, education, and length of residence.
[201] Cf. Herstatt & Hippel 1992, p. 218; Franke & Shah 2003, p. 158; Lüthje 2000, pp. 69f.
[202] Cf. Lüthje 2004, p. 693; Marchi et al. 2011, p. 351; Baldwin et al. 2006, p. 1296; Tietz et al. 2005, p. 336.
[203] Cf. Hienerth 2006, pp. 285f.; Marchi et al. 2011, p. 351; Füller et al. 2007, pp. 65 & 69; Jeppesen & Frederiksen 2006, pp. 55f.; Ogawa & Pongtanalert 2013, p. 44.
[204] Cf. Tietz et al. 2005, p. 336.
[205] Cf. Lettl et al. 2006, pp. 38f.

ahead of the market and they experience needs and trends much earlier than the average user. Hippel labeled them *lead users* and provided the following definition:

- *"Lead users face needs that will be general in a marketplace-but face them months or years before the bulk of that marketplace encounters them,* **and**
- *Lead users are positioned to benefit significantly by obtaining a solution to those needs."*[206]

The first component, *being ahead of trend*, is based on adoption processes as described in chapter 3.1 above. Some users adopt innovations much earlier than others and lead users pick up emerging market trends (e.g., products, tastes, technologies) first.[207] The second component, *high expected benefits*, assumes, that users who expect to profit from a specific innovation will be motivated to develop it themselves (see also chapter 3.2 above).

Lead users are typically highly qualified and very advanced users who are so far ahead of market trends that manufacturers have either not yet discovered their needs or have decided that development for this segment is not profitable due to its small size. Consequently, lead users rarely have the option to buy a product for their needs and rather must innovate themselves.[208] It is important to note that lead users do not represent a specialized niche with rare market demands, but that they actually are at the very forefront of the market and they anticipate (and possibly even create) relevant market trends.[209] Studies have shown that lead users adopt products and technologies approximately four to seven years before the market average.[210] Identifying lead users and incorporating them in product development and early product diffusion can, therefore, provide manufacturers with a competitive advantage.[211] Although both components of lead users are typically closely related in practice, they are conceptually independent. The high expected benefit is an indicator for innovation likelihood, while being ahead of trend indicates the potential commercial attractiveness of an innovation.[212]

To incorporate lead users into the product development process of manufacturers, a four-step process is proposed as follows: (1) Definition of lead user indicators (especially relevant trends and measures of potential benefits); (2) Identification of

---

[206] [Original emphasis in italics] Hippel 1986, p. 796.
[207] Cf. Rogers 1962; Hippel 1988, p. 107.
[208] Cf. Franke et al. 2006, p. 312; Herstatt et al. 2001, p. 2.
[209] Cf. Hippel 2007, p. 300.
[210] Cf. Urban & Hippel 1988, p. 573; Lüthje et al. 2002, p. 29.
[211] Cf. Schreier & Prügl 2008, p. 333.
[212] Cf. Franke et al. 2006, p. 311.

relevant lead users; (3) Creation of concept together with lead users (typically done in a workshop); (4) Testing of the concept with regular users.[213]

Exemplary lead users have been doctors in developing countries for innovations in surgical drapes at 3M, top athletes for innovations of sporting equipment, and disabled persons for innovations of age-based products.[214] Recent research suggests that some manufacturers do not have to invest many resources into the search for lead users because they can easily find them among their own employees as *embedded lead users*.[215]

The lead user method has proven to be very successful with industrial products. Manufacturers were able to create novel breakthrough products while decreasing development times and costs. The resulting products showed revenue potential eight times higher than regular products.[216] When the lead user method is applied to consumer goods there are some additional challenges. The number of users of consumer goods can reach millions and these users are mostly unknown. It is, therefore, very difficult to reliably and efficiently screen for and identify potential lead users.[217] Screening surveys over the internet have been applied to overcome this problem, but results have been unreliable.[218] Especially problematic for the case of consumer goods is that the lead user definition only provides two characteristics of suitable users, and the dichotomous separation of the population omits useful information.[219] It is therefore suggested that lead userness should be measured on a continuous scale to allow for more flexibility in different levels of lead userness. Morrison (1996) proposed the continuous construct *leading edge status (LES)*, which consists of the two lead user components *benefits recognized early* and *high level of benefits expected*, *perceived LES* (by self and by others), and actual *applications generation*.[220] This construct additionally takes into account actual innovative behavior and, therefore, provides an additional, easy-to-measure indicator Although the LES construct is rarely applied in user innovation research, many researchers refer to the term leading edge user and have agreed that lead userness is not a set of dichotomous characteristics but is rather something that should be measured on a continuous scale.

---

[213] Cf. Hippel 1986, p. 797; Urban & Hippel 1988, pp. 570ff.; Herstatt & Hippel 1992, pp. 214ff.
[214] See Hippel et al. 1999; Tinz 2007; Helminen 2008.
[215] See Schweisfurth 2013; Schweisfurth & Raasch 2012.
[216] Cf. Hippel et al. 1999, p. 56; Herstatt & Hippel 1992, pp. 219–220; Lilien et al. 2002, p. 1051.
[217] Cf. Schreier & Prügl 2008, p. 332; Ernst et al. 2004, p. 123.
[218] Cf. Tinz 2007, p. 97.
[219] Cf. Morrison et al. 1999, p. 5.
[220] Cf. Morrison 1996, pp. 13f.; Morrison et al. 2004, p. 356.

3.5 Antecedents of Lead User Characteristics and Innovative Behavior

Lead users are not necessarily user innovators. Additional characteristics and situational factors are also relevant.[221] Lüthje (2000), for example, enhanced the lead user concept and proposed to distinguish innovating *advanced customers* from non-innovating users with six characteristics: new needs, dissatisfaction, use experience, technical expertise, intrinsic motivation, and extrinsic motivation. Subsequently, researchers focused on these and some additional influencing factors. Table 2 below provides an overview of the analyzed influencing factors of innovativeness and lead userness in the most cited scientific contributions. Generally, there are two types of characteristics: highly context-specific ones (i.e., ahead of trend, expected benefits, use experience, product knowledge) and less context-specific ones (technical expertise, extrinsic and intrinsic motivators, innovativeness[222], and speed of adoption). Ahead of trend and new needs are usually used interchangeably, because users who are at the forefront of new trends experience new needs first. The same applies for high benefits and dissatisfaction: the more users are dissatisfied with existing products in the market, the higher the benefits they expect from a solution that would fulfill their needs.

Use experience, intrinsic motivations, and technical expertise are most often analyzed and show strong correlations with innovative behavior, as well as lead user components (see Table 2 below). Therefore, these three characteristics are especially well suited to act as indicators during the search for lead users.[223] Product knowledge and adoption behavior are less often analyzed but generally have a positive influence.[224] Extrinsic motivators rarely have any influence and seem to be only relevant for user-manufacturers. In some lead user workshops, lead users have even refused to accept payment, even when they were entitled to it, because they felt rewarded enough by being included in the development process.[225] Providing extrinsic motivators as a reward could even have the negative side effect of reducing intrinsic motivation.[226]

---

[221] Cf. Lettl et al. 2006, pp. 32f..
[222] In all cases not actual innovative behavior, but different innovativeness indices were applied, like the leading edge status Morrison et al. 2000, self-rated innovativeness Urban & Hippel 1988, domain-specific innovativeness Schreier et al. 2007, or Kirton's 1976 Adaptive versus Innovative Personality Inventory Schreier & Prügl 2008.
[223] Cf. Lettl et al. 2006, pp. 32f.
[224] Cf. Schreier & Prügl 2008, p. 342.
[225] Cf. Herstatt & Hippel 1992, p. 218.
[226] Cf. Franke & Shah 2003, pp. 173f.

Table 2: Overview of Studies Analysing Influencing Factors of Innovative Behavior and Lead User Components

Study	Dependent Variable	Ahead of Trend/ New needs	High Benefits/ Dissatisfaction	Use experience	Product knowledge	Technical expertise	Extrinsic motivator	Intrinsic motivator	Innovativeness index	Speed of adoption	Others
Shaw 1985	Commercial success of innovation										+ Interaction frequency
Voss 1985	Owner of innovation process					+	O				
Urban & Hippel 1988	Lead user components		+						+	+	
Vanderwerf 1990	Innovative behavior										+ Product tying possible
Herstatt & Hippel 1992	Lead user components						O	+			
Slaughter 1993	Dominating owner of innovation process			+		+	+				
Riggs & Hippel 1994	Incentives for user innovators						O	+			
Lüthje 2000	Innovative behavior	+	+	+	+	+	O	−			
Morrison et al. 2000	Innovative behavior	+				+	O	+	+		
Shah 2000	Innovative behavior	+	+	+			+	+			
Lüthje et al. 2002	Innovative behavior		+	+		+					
Franke & Shah 2003	Innovative behavior	+	+	+			O	+			+ Role in community
Lüthje 2004	Innovative behavior	+	+	+	+	+	O			+	
Morrison et al. 2004	Leading edge status	+	+								+ Application generation
Tietz et al. 2005	Innovative behavior			+	+	+		+			+ Tools, materials + Time
Hienerth 2006	Driving factors of user innovators	+	+				+	+			
Franke et al. 2006	Innovative behavior	+	+			+					+ Community resources
Schreier et al. 2007	Leading edge status[a]								+		+ Opinion leadership
Schreier & Prügl 2008	Lead userness			+	+				+	+	+ Locus of control
Marchi et al. 2011	Level of innovativeness				+			+			+ Strategic alignment
Schuhmacher & Kuester 2012	Idea Quality	O	+	O	O		O	+			O Involvement

+ Positive relationship
O No relationship
− Negative relationship

a) Leading edge status was actually the independent variable. Relationships were estimated in very simple SEMs so that relationships can also be interpreted in the other way

On top of influencing factors that apply to individuals only, the affiliation to a community can influence innovative behavior. An individual's role in a community and communication among members can facilitate the quantity and quality of innovations.[227] The innovation is then not necessarily the result of single individuals but of a collaborative process. The rise of the internet has provided the basis for the emergence of myriads of special-interest communities and, along with it, fast and immediate diffusion of information. Studies have shown the positive impact of communities on innovative behavior and the difference between individual and community innovators.[228] However, this research stream will not be part of this research project.

Although characteristics of lead users and user innovators have been identified and researched, most of them are still highly situation-specific. It would be helpful for the practical application of the lead user method if more general factors, like personality traits, could be identified.[229]

3.6 Development of User Innovation in Academic Research

The awareness that user innovation enhances the understanding of the product development process has recently gained a great deal of attention, as will be shown in the following analyses of contributions to the academic literature.

The following literature review is based on the Business Source Premier database, available via EBSCOhost[230]. The popular Google Scholar database was not chosen, because it includes a much broader range of sources (e.g., conference proceedings, book chapters, working papers), and does not exclusively contain scholarly journals. Also, it lacks some sources published before 1990.[231] Another popular database for citation analysis is Thomson Reuters' web of knowledge, but its coverage focuses mainly on US sources, and it only includes journals that are ISI-listed.[232]

The terms *user innovator* and *consumer innovator* are synonymously used in the literature and were both included for this analysis. Since it was irrelevant whether articles focused on *innovations* or the individual *innovators*, both terms were included in the search. The results were limited to peer-reviewed journals to include only

---

[227] Cf. Ogawa & Pongtanalert 2013, p. 42; Franke & Shah 2003, p. 164; Franke et al. 2006, p. 312.
[228] See for example Franke & Shah 2003; Füller et al. 2007; Janzik 2012.
[229] Cf. Franke et al. 2006, p. 313.
[230] http://search.ebscohost.com.
[231] Cf. Harzing & Wal 2008, p. 65. Website of Google Scholar: http://scholar.google.de.
[232] Cf. Harzing & Wal 2008, pp. 63f.. Website of Thomson Reuters' web of knowledge: http://www.webofknowledge.com.

qualitative academic results. The search string *"user innovat*"* OR *"consumer innovat*"* in all text fields resulted in a total of 789 articles, dating back to 1959.

As one can see in Figure 6 below, articles were only sporadically published until the late 1980s. From then on, interest in the topic of user innovation steadily increased until it rose quickly after 2000 to its hitherto peak in 2010.[233] The top 10 journals account for 35.4 % of all articles and come from the field of marketing sciences and technology and innovation management.[234]

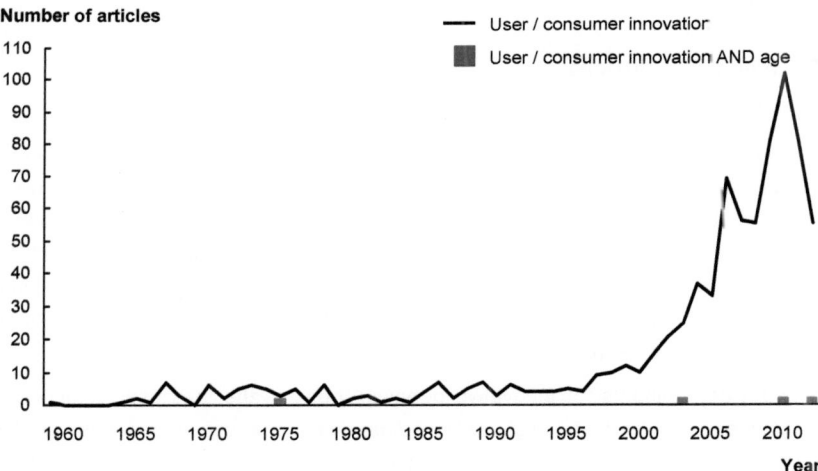

Figure 6: Development of Scientific Articles on User Innovation in Peer-Reviewed Publications from 1959 to 2012[235]

For the analysis of articles focused on user innovation and age, only articles in which age was of specific interest to the researcher (and not just one of many control variables) were relevant. Therefore, the search was limited to the abstracts and not all text fields. The search string *"user innovat*"* OR *"consumer innovat*"* AND age returned a total of 61 articles from 1975 to 2012. All abstracts were then screened to

---

[233] Extrapolation from published articles from January to April 2013 shows that full year figures should rise again to the level of 2011.
[234] Top 10 journals with most found articles (number of articles in brackets): 1st Advances in Consumer Research (45), 1st Journal of Marketing (45), 3rd Journal of Product Innovation Management (36), 4th International Journal of Innovation Management (31), 5th R&D Management (27), 6th Journal of Marketing Theory & Practice (21), 7th Journal of Consumer Research (20), 7th Journal of Marketing Research (20), 9th Journal of Marketing Management (17), 9th Management Science (17).
[235] Own illustration. In total 789 articles from 1959 to 2012 on Business Source Premier.

verify whether they actually dealt with age and were not false positives due to similar words, e.g., engage, usage, advantage, or average. In the end, only four articles remained, dating from 1975, 2003, 2010, and 2012 (see Figure 6 above). This demonstrates that this relationship has not received much attention in the academic world yet.[236]

These results show that, although the topic of user innovation has been of increasing importance for academic literature, the impact of the personal characteristics of innovators (especially age), has not yet been in the focus and still should be investigated further.

3.7 Interim Conclusions

The chapter has shown the relevance of user innovation for product development and marketing. The lead user method has proven to be an especially valuable tool to improve effectiveness and efficiency of new product developments. First nationwide, large N studies have shown that user innovation exists among the broad population and is not just limited to industrial goods or certain niche markets. Nevertheless, the focus of consumer goods research up until now has been on sports and leisure products, which specifically attract young users. The age of the users has not yet played a role in the definition of research designs.

Important determinants for innovative behavior, besides the lead user components (being ahead of trend and high expected benefits), are use experience, product knowledge, and technical expertise. Financial rewards play, if at all, only an inferior motivational role for user innovators. This is also shown in the free revealing of information regarding new developments, which is typical for user innovators.

---

[236] The first paper Green & Langeard 1975 follows the outdated view, that an innovative consumer is one that adopts products very early. The source of the actual innovation is regarded to reside with the manufacturer. Nevertheless, the authors find that the most innovative consumers for grocery products in France were to be found in the age range from 35 to 49. The second paper Im et al. 2003 finds that age is negatively correlated with consumer innovativeness, but also measures innovativeness by the adoption of new products. The third article Morrison & McMillan 2010 looked at the impact of user characteristics on the creation of user generated content. The only significant relationship regarding age was that older users are less likely to be involved in social networking sites. Finally, Hippel et al. 2012 analyzed consumer innovations in a representative study of British households. They found that innovators existed among all age groups above 18 years of age and that there was no significant relationship between age class and share of innovating consumers.

4 Research Questions and Hypotheses

4.1 Research Gap and Research Questions

Chapters 2 and 3 provided an overview on the current state of research on the SiMa and user innovation. Both phenomena are currently intensively studied but are not yet comprehensively understood.

Many SiMa studies in applied research exist (e.g., Sentha, Open ISA, and SMILEY; see chapter 2.2.2 above) that try to explain how product development for the SiMa should be implemented. Academic research focuses more on possibilities to segment older consumers (e.g., according to health, financial status, need for autonomy, or preferences) and the resulting impact on marketing strategies. These concepts refer only to the development of age-based innovations *for* and occasionally *with* the elderly; never *by* them.

Research on user innovation studies so far has always focused on product categories that are quickly growing and changing (e.g., emerging new sport activities, high-tech industries), and which are dominated by young users. Aging influences the capabilities and attitudes of individuals, but whether this translates to changes in innovative behavior is not yet known. A study that specifically analyzes the impact of age on user innovation does not exist. In a review of the current state of user innovation research, Bogers, Afuah, and Bastian (2010) call out for studies that "*[...] explore how the cognitive limitations [...] of economic actors affect their decision-making capabilities in the process of innovation.*"[237] Astor (2000, p. 322) specifically points out that there exist no empirical findings on the impact of age on the participants in the innovation process past the retirement age, and Sudbury and Simcock (2009) highlight "*[...] that there is a lack of valid and reliable empirical research available to help guide marketing strategies*"[238] for the SiMa.

In light of the fast growing SiMa and the fact that user innovations create social welfare by reducing deadweight loss[239], understanding SiMa user innovators could significantly contribute to the development of urgently required age-based innovations.

This research study tries to close this research gap and analyze how older users innovate, in case they do, and how user innovation changes with age. Therefore, the following four research questions were defined:

---

[237] Bogers et al. 2010, p. 866.
[238] Sudbury & Simcock 2009, p. 23.
[239] Cf. Hippel et al. 2012, p. 1678.

RQ1: Do user innovators exist in the Silver Market population?

RQ2: Which determinants of innovative behavior characterize the Silver Market user innovator? Do these determinants differ compared to younger user innovators?

RQ3: How strong - if there is one - is the moderating influence of chronological / cognitive age on the determinants of innovative behavior?

RQ4: Do user innovations by Silver Market user innovators differ from "regular" user innovations, and if so, how?

4.2 Hypotheses Regarding Silver Market User Innovators

The following hypotheses relate to the influencing factors of lead userness and innovative behavior by users. Together they form the structural model shown in Figure 7 below. The hypotheses of chapter 4.2.5 regarding the difference between the age groups are formulated under the assumptions that RQ1 is answered positively and that user innovators in the SiMa exist.

4.2.1 Use Experience

Frequent and repeated use of products or services leads to use experience. Schreier and Prügl (2008) define use experience as "*[...] learning from experience and [...] performance-related knowledge from primary product usage.*"[240] According to this definition, use experience requires time to accrue, and it can only be built up from directly using and interacting with a product. This primary product usage is required to familiarize oneself with the product. The formation of personal wants and needs in a certain domain is heavily correlated with the consumption of that domain's products.[241] Thereby, the experienced user can better identify and describe existing problems and analyze potential issues that might arise in the context of using it with other products or in divergent use scenarios. Through frequent usage, a user might also be able to conceive potential solutions for issues and test them in practice.[242]

Use experience has often been the focus of studies of influence factors of lead userness and innovativeness.[243] Users need to build up extensive knowledge about a

---

[240] Schreier & Prügl 2008, p. 336.
[241] Cf. Bünstorf 2003, p. 58.
[242] Cf. Lüthje 2004, p. 686.
[243] Cf. Slaughter 1993; Lüthje et al. 2005; Franke & Shah 2003; Tietz et al. 2005; Lüthje 2004; Schreier & Prügl 2008.

product and product-related tasks before they can extend the boundaries of these tasks. Based on this assumption, Schreier and Prügl (2008) argue that high levels of experience are a prerequisite for a high level of *lead userness*.[244] They also show that a person's use experience significantly influences a person's lead userness.[245] Use experience was even more strongly related to lead userness than a person's locus of control or innate innovativeness.

Use experience can be split into frequency of use, the overall time span that has elapsed since the first exposure (duration), and different specialties of a specific usage domain. The latter is more important for sports activities, but the first two are easily quantifiable and may be used for this research. Frequency and duration apparently positively influence the creation of ideas.[246] User innovators primarily draw their need information from their own personal experiences, rather than from information from others.[247] Few studies have found contrary evidence suggesting that use experience is not required to be innovative.[248] All of these studies were conducted in software application development. One can assume that use experience regarding IT support systems is not as important as use experience regarding the actual process in question.

To create ideas for new and improved products, personal use experience is a clear requirement. Based on the reasoning above, the following hypotheses were formulated:

    H1a: *Use experience is positively related to being ahead of trend.*

    H1b: *Use experience is positively related to high expected benefits.*

    H1c: *Use experience is positively related to innovative behavior.*

    H1d: *Use experience is positively related to product knowledge.*

4.2.2 Product Knowledge

Product knowledge "*[...] consists of know-how about the product architecture and the used materials and technologies of the existing products in the market.*"[249] A full understanding of the products available is required to identify blank spots that leave

---

[244] Cf. Schreier & Prügl 2008, p. 336.
[245] Cf. Schreier & Prügl 2008, p. 342.
[246] Cf. Lüthje et al. 2005, pp. 959f.; Tietz et al. 2005, p. 331.
[247] Cf. Hippel 2005a, p. 74.
[248] Cf. Feld 1990, p. 13; Voss 1985, p. 117.
[249] Lüthje 2004, p. 686.

room for improvement and innovation. It is also needed to translate tacit knowledge[250] on needs and requirements into concrete product specifications.

Through a clearer understanding of the limiting factors within their equipment and the required specifications for an optimal product, users with a high level of product knowledge should also be able to more precisely assess the expected benefits of an improvement.

Quantitative studies in online consumer communities have shown that a user's product knowledge positively affects an individual's *innovative behavior* (case: online brand community for motorcycles)[251] and that at the core of innovative communities (case: user designs for basketball shoes) there are members with extensive product knowledge[252]. Tietz et al. (2005) could show that this positive relationship also holds true for users in a physical consumer goods market.[253]

Based on the reasoning above, the following hypotheses were formulated:

    H2a: *Product knowledge is positively related to being ahead of trend.*

    H2b: *Product knowledge is positively related to high expected benefits.*

    H2c: *Product knowledge is positively related to innovative behavior.*

4.2.3 Technical Expertise

Technical expertise refers to knowledge regarding the architecture of products and engineering techniques required to actually build and modify products.[254] This knowledge is not necessarily domain-specific, and it is assumed that individuals with technical expertise can apply this knowledge to different problems. Technical expertise is required to transform a plain innovative idea into a working prototype. It, therefore, can explain why some users develop promising new products and prototypes while others stop at the idea stage.[255] Lettl, Herstatt, and Gemünden (2006) even argue that technical expertise accounts for the difference between an active development contribution in a limited user domain versus in a widely applicable technological domain.[256] It is important to distinguish between the technical knowledge of an individual and the technical resources one might have at

---

[250] Cf. Davenport & Prusak 1998, p. 95; Nonaka & Takeuchi 1995, pp. 8ff.
[251] Cf. Marchi et al. 2011, p. 354.
[252] Cf. Füller et al. 2007, p. 69.
[253] Cf. Tietz et al. 2005, p. 331.
[254] Cf. Franke et al. 2006, p. 307.
[255] Cf. Lüthje et al. 2005, p. 961; Lettl & Gemünden 2005, p. 343.
[256] Cf. Lettl et al. 2006, p. 37.

hand.[257] Limited access to technical resources might prevent an innovator from actually building a running prototype, but with enough individual technical knowledge the individual would already know how to build it. In the presented cases, the focus was solely on individual knowledge and regarded ideas and blueprints already as innovations (see chapter 7.1.2.1 below).

Some authors also noted that a certain lack of technical expertise can be overcome through motivation and high endurance to determine a possible solution.[258] In these cases, a trial-and-error approach built up the individual knowledge and finally led to a working solution.

Morrison, Roberts, and Hippel (2000) have proven the importance of technical knowledge for innovations in a B2B context among libraries in Australia and show that technical expertise correlates with lead user characteristics.[259] Although it affects both lead user components significantly, technical expertise seems to influence the notion of being *ahead of trend* more than the notion of *high expected benefits*.[260] Lüthje (2004) and Lüthje, Herstatt, and Hippel (2005) show that in cases of outdoor sporting equipment (climbing, cross-country skiing, and mountain biking) technical expertise is correlated with a deeper understanding of how the specific equipment functions, which is a prerequisite to use it to be ahead of the trend.[261] They also show that higher levels of technical expertise are related to having ideas for improvement. Franke and Hippel (2003) prove in a sample of IT software that technically skilled users were more satisfied with a system they modified than less technically skilled users. Based on the knowledge of their skills, these technically skilled users expected higher benefits before and then capitalized these benefits through their modifications.[262]

In one of the first cross-cultural consumer innovation studies, Hippel, Ogawa, and Jong (2011) demonstrate that technically trained individuals were much more likely to innovate than the average population.[263]

Based on the reasoning above, the following hypotheses were formulated:

> H3a: *Technical expertise is positively related to being ahead of trend.*

---

[257] Cf. Tietz et al. 2005, pp. 334ff.; Lettl et al. 2006, p. 36.
[258] Cf. Tietz et al. 2005, p. 336; Voss 1985, p. 117.
[259] Cf. Morrison et al. 2000, p. 1522.
[260] Cf. Franke et al. 2006, pp. 307f.
[261] Cf. Lüthje et al. 2005, pp. 961f.; Lüthje 2004, p. 691.
[262] Cf. Franke & Hippel 2003.
[263] Cf. Hippel et al. 2011, p. 31. The study collected data in the United States, United Kingdom, and Japan. The differences innovative behavior between technically trained individuals and the average population were especially visible in the two Western countries (share of innovators was between +54 % and +97 %). The difference was much lower (+13 %) in Japan.

H3b: Technical expertise is positively related to high expected benefits.

H3c: Technical expertise is positively related to innovative behavior.

H3d: Technical expertise is positively related to product knowledge.

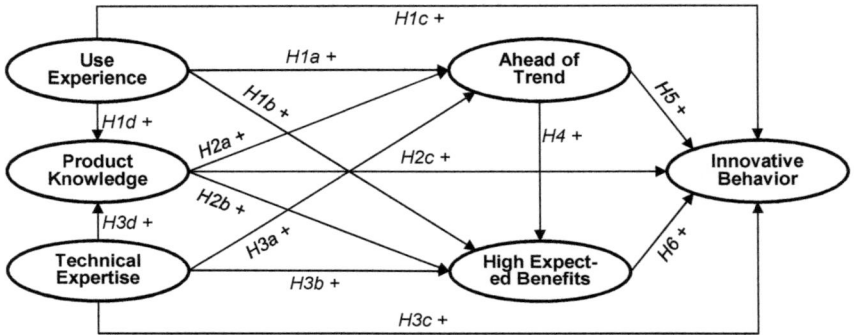

Figure 7: Overview of Hypotheses Regarding Silver Market User Innovators without Moderating Influence of Age[264]

4.2.4 Lead User Characteristics

Users with lead user characteristics possess needs that average users have not yet experienced, and they are motivated to find solutions for these needs because they can benefit significantly from them.[265] Researchers have often regarded lead userness and its two defining factors as one singular construct and have only analyzed the relationship of lead userness on innovativeness, opinion leadership, etc.[266] Franke, Hippel, and Schreier (2006) argue that the two components *ahead of trend* and *high expected benefits* are conceptually independent and are not necessarily related in every case.[267] They could show that in the case of extreme sports both components, although related, stimulated different innovation characteristics. While both components increased the innovation likelihood, being *ahead of trend* led to more attractive innovations. This study followed the argument of Franke, Hippel, and Schreier (2006) and always included both components of lead

---

[264] Own illustration.
[265] Cf. Hippel 1988, p. 107; Herstatt et al. 2001, p. 2.
[266] Cf. Urban & Hippel 1988; Lilien et al. 2002; Schreier et al. 2007; Schweisfurth 2013; Hippel 1986; Herstatt & Hippel 1992.
[267] Cf. Franke et al. 2006, p. 303.

userness separately in the analysis. Since both components independently influence innovative behavior, they do not affect each other's impact on innovative behavior.

There exists sufficient evidence in the literature that lead userness has a positive impact on innovative behavior, independent of industry or product type. Lead users adopt new products earlier (Urban and Hippel (1988) showed an average of seven years for B2B products) and in a greater number.[268] That lead users are a superior source for novel innovations was shown for the industry conglomerate 3M[269], IT software[270], and sports equipment[271]. They generate innovations faster[272] and their results are more attractive[273].

Based on the reasoning above, the following hypotheses were formulated:

H4: *The lead user component being ahead of trend strongly positively impacts the lead user component high expected benefits.*

H5: *Being ahead of trend is positively related with innovative behavior.*

H6: *High expected benefits are positively related with innovative behavior.*

H7: *High expected benefits do not mediate the relationship between ahead of trend and innovative behavior.*

4.2.5 Moderating Influence of Age

As outlined in chapters 2.5 and 3.6 above, literature on the relationship between age and innovative behavior is very limited. The few studies that exist have mostly analyzed the impact of age on inventive output of individuals.[274] Verworn, Schwarz, and Herstatt (2009) have shown how to adapt HRM strategies to mitigate the effects of changing workforce demographics. Research on the specific moderating influence of age on the antecedents of innovative behavior simply does not exist at all. Gwinner and Stephens (2001) note that the literature on cognitive age analyzes antecedents and consequences of cognitive age and interprets it as a mediator

---

[268] Cf. Urban & Hippel 1988, p. 573; Schreier & Prügl 2008, p. 342; Schreier et al. 2007, p. 26.
[269] Cf. Lilien et al. 2002, p. 1051.
[270] Cf. Olson & Bakke 2001; Urban & Hippel 1988; Morrison et al. 2000.
[271] Cf. Lüthje 2004; Schreier & Prügl 2008; Hienerth 2006.
[272] Cf. Schreier & Prügl 2008, p. 334.
[273] Cf. Schreier & Prügl 2008, p. 334; Franke et al. 2006; Schuhmacher & Kuester 2012, p. 436.
[274] See Hippel et al. 2012; Ogawa & Pongtanalert 2011; Eisfeldt 2009; Oberg 1960; Adenauer 2002.

variable; *"however, this mediated relationship has never been subjected to an empirical test."*[275]

Since this research stream is still in a very exploratory state, interviews with academics in the field were conducted to discuss the potential moderating impact of chronological age on the relationship of the determinants of lead userness and innovative behavior, to derive the hypotheses. Six experts whose research focus was either on user innovation, product-development or age-related research provided their input.[276] Although it is uncommon to base hypotheses on expert interviews, there exist articles which have done so.[277]

Table 3: Expert Evaluations of Moderating Impact of Age

Impact of age on the relationship...	Positive	Neutral	Negative
... use experience → ahead of trend	17 %	17 %	67 %
... use experience → high expected benefits	33 %	17 %	50 %
... use experience → innovative behavior	17 %	17 %	67 %
... product knowledge → ahead of trend	33 %	17 %	50 %
... product knowledge → high expected benefits	33 %	17 %	50 %
... product knowledge → innovative behavior	17 %	17 %	67 %
... technical expertise → ahead of trend	-	83 %	17 %
... technical expertise → high expected benefits	17 %	67 %	17 %
... technical expertise → innovative behavior	-	83 %	17 %
... ahead of trend → high expected benefits	80 %	20 %	-
... ahead of trend → innovative behavior	60 %	40 %	-
... high expected benefits → innovative behavior	40 %	40 %	20 %

Table 3 above provides a summary of the expert evaluations of twelve statements. As one can see, there is no statement upon which experts voted consentaneously. Most experts noted that the prediction of the impact is very difficult. This explains the variation in the responses. It is also striking that in only three out of the twelve relationships, no impact was expected by the majority of the experts, although they were specifically instructed that the "neutral"-option was an acceptable response. This could be an indication that experts were influenced by the Hawthorne effect and were trying to provide meaningful answers.[278]

The impact of use experience on the two components of lead userness and innovative behavior was mainly expected to be affected negatively by age. Since use experience accumulates almost naturally with age, a relative advantage of greater user experience only exists at younger ages. In an older age group, a high degree of

---

[275] Gwinner & Stephens 2001, p. 1033. Although moderator and mediator effects are not the same, the statement correctly represents the state of the literature, that explicit interaction effects are not empirically tested.

[276] The list of experts and their field of expertise can be found in the lower part of Appendix 1.

[277] See for example Shepherd et al. 2011, published in the Academy of Management Journal.

[278] See Adair 1984.

experience does not present a competitive advantage and, therefore, should be less associated with being ahead of trend.[279] Additionally, older users typically value security over risk-taking, reducing their desire to be ahead of trend.[280] Functional fixedness occurs when people are so familiar with a product that they can hardly imagine a different way of using it or find alternative products for the same purpose.[281] Since the cognitive capacity and the fluid intelligence of older users is lower, functional fixedness probably presents a larger hurdle to become a lead user or innovator for them.

Based on the reasoning above, the following hypotheses were formulated:

H8a: Age negatively moderates the impact of use experience on ahead of trend.

H8b: Age negatively moderates the impact of use experience on high expected benefits.

H8c: Age negatively moderates the impact of use experience on innovative behavior.

The evaluations of the moderating impact of age for the relationships based on product knowledge were not as explicit as in the case of use experience but were still assignable. The experts mainly pointed out that product knowledge has a "*decreasing incremental effect*"[282] and product knowledge might be more prone to obsolescence than use experience. Therefore, the additional gain through an increase in product knowledge declines so that older people, who have accumulated a great deal of product knowledge over time, benefit less from new knowledge than younger people. In contrast, product knowledge can become outdated and therefore worthless if new products, technologies, and techniques are introduced to a market. Since older consumers tend to rely more on recommendations by family members and friends when making purchasing decisions, they put less trust in their own product knowledge.[283]

Based on the reasoning above, the following hypotheses were formulated:

H9a: Age negatively moderates the impact of product knowledge on ahead of trend.

---

[279] Supported by expert interview #1.
[280] Cf. Sudbury & Simcock 2009, p. 30; Dychtwald & Flower 1990. Also mentioned in expert interview #6.
[281] Cf. Adamson 1952, p. 288; Fichter 2005, p. 358.
[282] Expert interview #5.
[283] Cf. Moschis 1992b, pp. 259f.

H9b: Age negatively moderates the impact of product knowledge on high expected benefits.

H9c: Age negatively moderates the impact of product knowledge on innovative behavior.

Just like in the case of product knowledge, the experts pointed out that technical expertise is affected by obsolescence over time, although to a lesser degree.[284] Technical expertise was typically associated with an engineering background. This background would remain consistent regardless of age and would not influence the relative trend position of an individual.[285] Additionally, Becker (2000) argues for a formative period between the ages of 10 and 25. *"The formative period is also a phase in the life course that requires the acquisition of important values and norms, which usually stay with an individual for a long time, although they may be modified or reinforced later in life by further societal changes. The formative period is, furthermore, a phase in life in which individuals acquire a lot of skills."*[286] The moderating influence of age on the impact of technical expertise should, therefore, be relatively low, because it is mostly determined during early adulthood and only marginally changes thereafter.

When looking for new products, the elderly especially focus on comfort and convenience.[287] New functionalities and technical sophistication are less important buying criteria. Older users would therefore probably expect fewer benefits based on their technical expertise. Instead, they would focus on benefits that they can derive from their need for autonomy.[288] As outlined in chapter 4.2.3 above, technical expertise is important in realizing an idea and transforming it into a working prototype. A lack of it can be overcome through motivation and endurance.[289] Older users, especially those who are already retired, typically have more time available and are, therefore, in a position to invest the time needed to overcome a certain initial deficit in technical expertise.[290]

Despite potential arguments for a negative moderating influence, the majority of experts stated that they do not believe that the impact of technical expertise is

---

[284] Supported by expert interviews #1 and #3.
[285] Supported by expert interview #1 and #5.
[286] Becker 2000, pp. 115f.
[287] Cf. Wolfe 1994; Arnold & Krancioch 2011, p. 155.
[288] Cf. Kohlbacher et al. 2011b, pp. 4ff.
[289] Cf. Tietz et al. 2005, p. 336; Voss 1985, p. 117.
[290] Supported by expert interview #2.

significantly negatively moderated by age. Instead they argued that technical expertise and its influence remains largely *"stable with age."*[291]

Based on the reasoning above, the following hypotheses were formulated:

> *H10a: Age does not moderate the impact of technical expertise on being ahead of trend.*
>
> *H10b: Age does not moderate the impact of technical expertise on high expected benefits.*
>
> *H10c: Age does not moderate the impact of technical expertise on innovative behavior.*

Older users have different needs, caused partly by a decline in cognitive capacity and physical strength. Therefore, they require products that specifically respond to their needs, e.g., effortless gardening tools, easy-to-use pillboxes, or supporting bath lifts. Regular products that do not respect these needs will quickly generate dissatisfaction among older users.[292] If these older users are more active and possess an advanced trend position, they will push the limits of these products even faster and recognize inadequacies earlier.[293] As SiA's favor security and reliability when selecting products, they are less likely to be ahead of trend than younger users.[294] They also have less access to information about new trends and technologies (mainly because they use social media less). Under these assumptions, the likelihood of being ahead of trend is much rarer among older users. In other words the difference between individuals ahead of trend and not ahead of trend is larger between older users, and the individual will most probably experience more dissatisfaction with existing products and be more motivated to develop improvements.[295] As discussed above, older users also generally have more free time. In combination with high degrees of the lead user characteristics this could lead to more innovative behavior, as they can invest more time in experimenting and thinking about new ideas. One expert especially highlighted that the ability to transform one's expertise into innovative behavior is *"[...] associated with one's*

---

[291] Expert interview #6.

[292] Products for older consumers do not necessarily need to be designed specifically for the elderly as long as they take their requirements into account as well. The principles of universal design (see chapter 2.2.2) are a good example for how to design products while paying attention to all potential consumer needs.

[293] Supported by expert interviews #4.

[294] Sudbury & Simcock 2009, p. 31; Simcock et al. 2006, pp. 359ff.

[295] Supported by expert interview #2 and #4.

cognitive capabilities"[296], i.e., that in the case of older innovators, this is especially relevant in a low-tech environment.

Based on the reasoning above, the following hypotheses were formulated:

> *H11: Age positively moderates the impact of ahead of trend on high expected benefits.*
>
> *H12: Age positively moderates the impact of ahead of trend on innovative behavior.*
>
> *H13: Age positively moderates the impact of high expected benefits on innovative behavior.*

4.3 Propositions Regarding Innovation Characteristics of Silver Market User Innovators

As described in the previous chapter, research studies on the impact of age on innovative behavior and its antecedents do not exist, and even experts in the field of user innovation and SiMa have difficulty agreeing on the anticipated impact of age. Due to the lack of confirmed findings, the analysis of the characteristics of innovations, as well as the innovation process regarding differences between age groups, is therefore very exploratory. Instead of deriving hypotheses, propositions were formulated: an approach that reflects the current state of research better.

The difference between propositions and hypotheses is often not clear because researchers use these labels interchangeably.[297] For this research, the approach of Bailey (1994) was followed and hypotheses were only formulated for relationships that are directly testable and are therefore falsifiable.[298] Propositions represent a prior step in the thinking and discuss suggested relationships based on logical thinking as well as qualitative and descriptive data.

---

[296] Expert interview #1.
[297] Cf. Hage 1994, p. 100.
[298] Cf. Bailey 1994, pp. 43f.

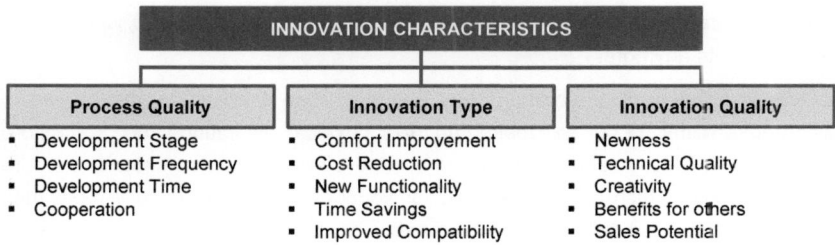

Figure 8: Dimensions of Innovation Characteristics[299]

The characteristics of the innovations in the research sample were divided and analyzed along three dimensions: process qualities, innovation type, and innovation qualities (see Figure 8 above).

Process Qualities

The process qualities contained the furthest development stage of the latest innovation, the time required to reach that stage, the frequency of innovation with regards to products owned, and the cooperation with others during ideation and realization.

According to Lettl, Herstatt, and Gemünden (2006) the actual development of new products requires imagination capabilities, a high level of expertise in the domain, tolerance of ambiguity, and technological expertise.[300] Older users typically have higher expertise in their user domains, because they have had much more time to be active in them. Conversely, technological expertise degenerates over time because of new standards, technologies, and techniques. It is therefore assumed that technological expertise decreases with age unless it is constantly refreshed. Studies on the output of scientists and engineers have also shown that creativity slowly decreases with age.[301] These studies have also shown that older employees can typically compensate for a lack in creativity with more relevant experience and increased social competences. One of the most relevant barriers to successful user innovation (after technological complexity) is time constraints.[302] Older people, especially after their retirement, have more leisure time available, and time is usually

---

[299] Own illustration.
[300] Cf. Lettl et al. 2006, p. 39.
[301] Cf. Hoisl 2007, p. 21; Oberg 1960, pp. 251ff.; Eisfeldt 2009, p. 166; Bergmann et al. 2006, p. 19.
[302] Cf. Braun & Herstatt 2009, p. 94; Tietz et al. 2005, p. 334.

not a scarce resource anymore.[303] Therefore, they should not need to worry about time constraints anymore.

The effects of aging on the human body and mind (see chapter 2.4 above) lead to new requirements, e.g., to make up for physical limitations or to restore independence. Older users could therefore more often experience situations where regular products, which were developed for younger users, do not fully meet their requirements. If that holds true, they probably have more ideas regarding how to improve and adapt existing products.

The size of the social networks of the elderly is smaller, and their number of social contacts decreases with age.[304] At the same time the pursuit of reputation and social connectedness becomes more important, creating an opposing effect.[305] Elderly people want to achieve something and have a feeling of purpose, especially after they have retired from professional life.[306] Therefore, they might cooperate more with others during idea generation and development of new products than younger innovators.

Based on these statements, the first proposition, regarding the process qualities of SiMa user innovations, was formulated as follows:

> P1: *The innovation process will differ between Silver Market user innovators and younger user innovators, especially with regard to development stage, frequency, and cooperation during development.*

Innovation Type

Innovations can be categorized in many ways. Often, the differentiation into incremental versus radical innovation is in the focus of research studies.[307] Some authors have argued that user innovations are typically medium-innovative and are only rarely radical.[308] This study, therefore, focuses only on the underlying purpose for which the innovation was developed. The potential purposes have been

---

[303] Cf. Lumpkin et al. 1989, p. 178.
[304] Cf. Backes & Clemens 2008, p. 75.
[305] Cf. Schewe 1991, p. 63.
[306] Cf. Schewe 1991, pp. 62f.. The search for a purpose can be witnessed in the growing number of educational programs at universities and specialized institutes as well as elderly volunteer organizations (e.g., Senior Corps, Age UK, iTNAmerica, and Freiwilligendienste aller Generationen).
[307] Cf. Lettl et al. 2006, p. 28; Fichter 2005, p. 357.
[308] Cf. Lettl et al. 2006, p. 29.

discussed with industry experts[309], and the five most often mentioned innovation types have been selected: new functionality, comfort improvements, cost reduction, time savings, and improved compatibility.

As was mentioned above, the needs and requirements of older consumers are different than those of younger ones. It can therefore be assumed that also the innovation type of products developed by older user innovators will be different.

Security and safety, as well a need for autonomy, are important values for older consumers.[310] The analysis of risk evaluations before purchasing decisions has shown that older consumers evaluate physical risk more than younger consumers. Time and financial risks, on the other hand, do not play a significant role.[311] It can therefore be assumed that innovations aimed at comfort and safety play a stronger role than innovations that improve the efficient use of time and cost, which are not as relevant for older consumers. Wolfe (1994) recommends marketing products for the elderly that emphasize comfort and the resulting experience of its use. It is assumed that this recommendation can be transferred to the innovations that older users develop.

Based on these statements, the second proposition, regarding the innovation types, was formulated as follows:

> P2: *Silver Market user innovators will focus on different innovation types, e.g., more on comfort and compatibility and less on time and cost reduction.*

Innovation Quality

User innovations can differ concerning their innovation qualities. Building on previous studies of innovation qualities, the focus is on the categories newness, technical quality, creativity, benefits to others, and sales potential.[312]

As described in the previous sub-chapter, differences in the values and risk evaluations of older consumer could lead to different innovation types. The same reasoning applies to innovation qualities. Older users are typically less interested in products that offer completely new functionalities. Instead, they want to maintain and respectively regain their independence and perform activities autonomously.[313]

---

[309] Consulted experts were Mr. Lemke, Mrs. Leipelt, and Mr. Gröll – all very knowledgable in the field of the German camping industry. See Appendix 1 for details.
[310] Cf. Dychtwald & Flower 1990; Kohlbacher et al. 2011b, p. 5; Kohlbacher et al. [in press].
[311] Cf. Simcock et al. 2006, pp. 357ff. & 365.
[312] See Lüthje et al. 2002; Franke & Shah 2003; Franke et al. 2006.
[313] Cf. Kohlbacher et al. 2011b, p. 5; Randers & Mattiasson 2004, p. 69.

Regular user innovators evaluate their own innovations as only moderately new.[314] Older users probably put even less emphasis on the newness of their products, as long as they serve their needs.

Although it is just a stereotype that older consumers avoid new technologies, it has been shown that the adoption of high technology products is less common among older consumers, especially if they are older than 65.[315] They may, therefore, also put less emphasis on the technical sophistication of their own innovations. As was shown in detail in chapter 2.5 above, creativity declines with age. This will most probably also apply to the resulting user innovations.

Due to the demographic shift, the SiMa is currently growing. Age-specific or universal products do not yet exist for all product categories, and the industry is only slowly picking up this trend.[316] The elderly also show much compassion for others and can empathize with other older people's needs.[317] Accordingly, the potential market size for products that solve age-specific needs is rather large, and many other users could potentially benefit from tailored product innovations. The resulting sales potential in case a user innovation is commercialized is, therefore, high. A contrary effect is the discrimination of age-based products in the retail sector[318], but it is doubtful that users are aware of this.

Based on these statements, the third proposition, regarding the innovation qualities of SiMa user innovations, was formulated as follows:

> P3: *Innovations by Silver Market user innovators will exhibit different qualities than those by younger user innovators. They are likely to score lower on newness, technical quality, and creativity but higher on benefits to others and sales potential.*

Age Difference of Cognitive versus Chronological Age

The use of chronological age for research is very limited because it is not strongly related to the behavioral and attitudinal patterns of older people.[319] Cognitive age is based on the self-perception of individuals, and it incorporates their evaluation of their health and financial status, social network, and capabilities (see chapter 2.3.2

---

[314] Cf. Lüthje et al. 2002, p. 16; Franke & Shah 2003, p. 163.
[315] Cf. Moschis 1992b, pp. 276f.; Fisk et al. 2009, p. 5.
[316] Cf. Kohlbacher & Herstatt 2008a, p. xi.
[317] Cf. Plutzer & Berkman 2005, p. 80.
[318] Cf. Levsen 2015. The diffusion of age-based innovations is often blocked by retailers who will not allow these products to receive sufficient attractive shelf space. Apparently retailers feel that negative emotions towards age-based products could spill over on their own image or other products in store.
[319] Cf. Barak & Schiffman 1981, p. 602.

above for more details). It is coherent with the self-concept theory, which argues that a person's self-concept is a function of behavior effects, and that it is related to individual consumer behavior.[320] The age difference of cognitive age and chronological age particularly indicates how an individual evaluates himself in comparison to his age cohort. Research on older users based on cognitive age has uncovered several differences. Cognitively younger users show a different ranking of values[321], a more active life-style orientation[322], higher information seeking behavior and less cautiousness in purchases[323], and higher innovativeness[324]. A study by Szmigin and Carrigan (2000) could not confirm differences in cognitive age between groups of high and low consumer innovativeness, but they failed to report figures on the age differences of their sample.[325]

Since the age difference indicates differences in personal values and consumer behavior, it can be assumed that the resulting innovations of user innovators will differ according to the size of the age difference.

Based on these statements, the fourth proposition, regarding the impact of the age difference on the characteristics of innovations by SiMa user innovators, was formulated as follows:

> P4: Among Silver Market user innovators, cognitively younger user innovators will exhibit differences related to process quality, innovation type, and innovation quality compared to cognitively older user innovators.

---

[320] Cf. Sirgy 1982, pp. 291ff.
[321] Cf. Kohlbacher & Chéron 2011, p. 183; Sudbury & Simcock 2009, p. 31. In the study by Sudbury & Simcock 2009, out of eight values based on the list of values (LOV) by Kahle 1983, only self-respect showed consistent importance across all cognitive age groups. All other values showed partly very strong differences.
[322] Cf. Wilkes 1992, p. 299.
[323] Cf. Gwinner & Stephens 2001, p. 1044; Stephens 1991, p. 45.
[324] Cf. Barak & Schiffman 1981, p. 603.
[325] Cf. Szmigin & Carrigan 2000, p. 518.

Part B. QUANTITATIVE EMPIRICAL STUDY

5 Introduction to the Research Field: Camping & Caravanning

5.1 Characterization of Camping Market

5.1.1 Origin and History of Camping

For thousands of years, humans have been sleeping outdoors in tents or temporary facilities. Our ancestors have lived in self-made tents, and many nomadic tribes (particularly in Central Asia) still continue to do so today. Modern camping is much younger and its roots date back to the late 19th century. As will be detailed in the following chapter, many major steps in the development of camping and caravanning, as well as camping itself, are based on user innovations.

The Britannica Encyclopedia defines camping as a *"recreational activity in which participants take up temporary residence in the outdoors, usually using tents or specially designed or adapted vehicles for shelter."*[326] The focus of camping is on recreational activities that individuals undertake because they want to and not because they have to (contrary, for example, to nomadic tribes who follow food and water sources).

Traditionally, carriages were used solely for transporting goods and people. Travelling artists and the Romani people were the first to also use them as their living quarters. The first *leisure trailer* was built by the Bristol Carriage Company for Dr. William Gordon Stables in 1885,[327] but it remained a unique specimen. A man who was also familiar with traveling long distances in horse-drawn carriages was Thomas Hiram Holding (1844 – 1930). He is considered to be the founder of modern recreational camping. Holding gathered his first experiences on traveling overland during a 1,900 km long journey through the prairies of America with his family in 1853.[328] In 1887, he traveled with a canoe through the highlands of Scotland, deciding to camp there. A few years later, he undertook the first bicycle camping trip, for which he invented some portable camping equipment himself.[329] Based on his experiences and innovations, Holding published two books: "Cycle and Camp" in

---

[326] Ryalls & Petri.
[327] Cf. The Caravan Club Limited 2012.
[328] Cf. Ryalls & Petri.
[329] Cf. Campinginfo.org 2012.

1898 and "The Camper's Handbook" in 1908.[330] After the first book was published, other camping enthusiasts contacted Holding, and in 1901 they founded the Association of Cycle Campers in the UK, which would later become "The Camping and Caravanning Club".[331] In 1933, 16 clubs from seven countries founded the "Fédération Internationale de Camping et de Caravanning" (F.I.C.C.).[332] In Germany, the first official camping association "Deutscher Camping-Club e.V." was founded on September 22, 1948; it remains to be the largest camping association in Germany.[333]

The camping community quickly grew in the early 20th century[334] but it was overshadowed by the consequences of World Wars I and II. People's leisure time and wealth increased with the economic revival that began in the 1950s and camping quickly grew to be one of the preferred leisure activities.

With the advent of affordable cars, the popularity of caravans increased. The first caravans were built by users who wanted to combine the comfort of horse-drawn carriages with the advantages of automobiles. Those caravans were made of used motor car parts, plywood, and canvas, and their design was often inspired by boats and their cabins.[335] The first caravan in Germany was probably built in 1934 by journalist Heinrich Hauser, who wanted to "*Reisen und dabei gleichzeitig zu Hause bleiben*" ("*travel while staying at home*"; translation by Konstantin Wellner)[336] with his family. Other camping enthusiasts also built their own caravans, and soon other people wanted to buy their models. Some of the inventors used that opportunity to become the first caravan manufacturers, e.g., Sportberger (inventor of the first pop-up caravans), Dethleffs, and Westfalia (the latter two are still in business today).[337] The number of caravans grew quickly: In 1954, there were 1,017 caravans officially registered in Germany. Just ten years later this number grew to 39,386, and today it is estimated to be around 900,000.[338]

---

[330] Cf. Holding 1898, 1908.
[331] For a detailed description of the development of the Camping and Caravanning Club, please refer to The Camping and Caravanning Club 2013. Holding was also the first president of The Camping and Caravanning Club. The adventurous and innovative character of the organization was also mirrored in its presidents. Among others, Robert Falcon Scott who belonged to the first 10 people to reach the South Pole and Sir Robert Baden-Powell, founder of the Boy Scouts, held the president post.
[332] Cf. Fédération Internationale de Camping 2013.
[333] Cf. Der Deutsche Camping-Club e.V. 2012.
[334] One of the first campsites was opened on the Isle of Man in 1894. By the end of the 1800s it attracted 600 people per week, in 1904 additional land was purchased to increase capacity for 1,500 tents. Cf. Campinginfo.org 2012.
[335] Cf. The Camping and Caravanning Club 2013.
[336] Hauser 1935, p. 7.
[337] Cf. Thünker 1999, pp. 67ff.
[338] Cf. Hierhammer 1997, p. 172; Caravaning Industrie Verband e.V. (CIVD) 2013b.

Westfalia also led the next leap in the evolution of caravans when they introduced the first motorized caravan, based on the newly introduced Volkswagen Transporter.[339] The *VW Bulli* quickly gained a considerable fan base and was a model for many more vehicles that would follow. According to estimates of the CIVD, there existed around 400,000 motor caravans in Germany and almost 1,000,000 additional units elsewhere in Europe by the end of 2012.[340] Users have continuously provided manufacturers with improvement ideas for camping vehicles and equipment. These suggestions range from small alterations to the layout of caravans to completely new products, like roof top tents.[341]

Several trends exist within the camping community that focus on very specific target groups. *Glamping* and *spa camping* are especially relevant for the elderly. Glamping ("glamorous camping") appeals to tourists who seek to combine the comfort and luxury of an upscale hotel with a nature experience and do not want to carry and maintain their own tents and equipment. Lodgings for this segment are typically semi-permanent tents such as yurts, tipis, and safari tents that include full beds, en suite bath rooms, and full board. Prices correspond to the high standard and can go up to several thousand Euros per night.[342] Therefore, glamping does not appeal to the regular camping tourist and attracts previously untapped customer segments.

Spa camping is a predominantly German phenomenon with specialized campsites in Southern Germany and along the German coastline. Some spa campsites also exist in Austria, Switzerland, Italy and France.[343] During spa camping, the regular camping vacation is combined with a stay at a health resort. The treatments are sometimes received on the campsite and sometimes in specialized medical spas. Spa campsites generally have a very high standard and focus on modern facilities and a comfortable stay.[344] This is especially appealing to elderly campers who can combine their required treatments with their favored way of traveling.

---

[339] Cf. Westfalia Mobil GmbH 2013.
[340] Cf. Caravaning Industrie Verband e.V. (CIVD) 2013b.
[341] One of the most famous examples for a roof-top tent is the "Villa Sachsenruh" that was invented by Gerhard Müller. This roof top tent was specifically built to fit on the Trabant, the most common car of the GDR. Due to the limited supply of hotel rooms as well as camping vehicles, this roof top tent provided an affordable alternative. Cf. Thünker 1999, pp. 45f. & 86f.
[342] Cf. Wikipedia contributors 2013b.
[343] For an overview on specialized campsites in Europe the search function of ADAC's online camping guide is a very helpful tool: http://campingfuehrer.adac.de/campingfuehrer/suche.php.
[344] In Germany they are usually classified as a 5***** campsite. An overview of 5***** campsites in Germany can be found here: http://www.camping-in-deutschland.de/campingplaetze/5-sterne/. A European overview can be selected here: http://en.camping.info/campsites?showTab=equip.

5.1.2 Camping in Germany and around the World

Camping in Germany is a popular leisure activity in which large parts of the population participate. According to the latest general stocktaking in Germany, which occurred in 2009, there exist 3,624 campsites with 286,985 camping spots for tourists, 347,090 for permanent campers, and 13,646 rented accommodations.[345] In total, there exist 647,721 camping spots. While the number of campsites has remained constant over the last few years, the structure of the camping spots has changed. The demand for permanent camping spots is decreasing[346] and campsites try to compensate by increasing the number of tourist spots and rental accommodations. Also, improvements in the campsites' infrastructure and additional facilities have led to a slight decrease in overall capacities.[347]

With 23 million overnight stays per year, Germany ranks fifth among the most visited countries by camping tourists. Leading by far is France with 98.8 million stays, followed by Italy (65.2 million), the United Kingdom (61.4 million), and Spain (31.1 million).[348] These five countries account for 76.7 % of all overnight stays in Europe.

Regarding the number of registered caravans and motor caravans, Germany is the leading market within Europe. According to estimations of the European Caravan Federation, 900,000 of the 4,054,900 European touring caravans and 440,000 of the 1,375,600 motor caravans are registered in Germany.[349] Some of the largest vehicle manufacturers also have their headquarters in Germany, e.g., Dethleffs GmbH & Co. KG, Fendt-Caravan GmbH, Hobby Wohnwagenwerk GmbH, Hymer AG, Knaus Tabbert GmbH, and Westfalia Mobil GmbH. 2012 was a record year for manufacturer of vehicles and accessories in Germany. Revenues for new vehicles reached 3.4 billion EUR while used vehicles for 2.3 billion EUR and accessories for 0.6 billion EUR were sold.[350] The market for camping equipment is very fragmented, and reliable information on the full market size is not available.

The total economic impact of camping tourists is also significant. The latest study with detailed data for Germany was conducted in 2010 by the Federal Ministry of Economics and Technology.[351] According to this study, camping tourists create a total of 11.6 billion EUR in net revenues each year. Expenses for vehicles and

---

[345] Cf. Bundesministerium für Wirtschaft und Technologie (BMWi) 2010, p. 9.
[346] Bundesministerium für Wirtschaft und Technologie (BMWi) 2010, p. 10 notes a drop of 7.6 % from 2003 to 2009.
[347] Cf. Bundesministerium für Wirtschaft und Technologie (BMWi) 2010, p. 10.
[348] Cf. Bundesministerium für Wirtschaft und Technologie (BMWi) 2010, p. 13. Figures from 2008.
[349] Cf. European Caravan Federation 2012b; European Caravan Federation 2012a.
[350] Cf. Caravaning Industrie Verband e.V. (CIVD) 2013a.
[351] See Bundesministerium für Wirtschaft und Technologie (BMWi) 2010.

equipment add up to 3.0 billion EUR per year. Permanent campers spent the least, with an average of 854 EUR, while tourist campers with a motor caravan spent more than 4,500 EUR annually on average.

The United States remains the largest market for camping. In 2011, 42.5 million people went camping, representing almost 15 % of the overall population above 6 years. Together they accumulated a total of 534.9 million overnight stays, nearly 50 % more than all overnight stays in Europe.[352] The main destinations for campers in the US are state and national parks, with their public campsites. Since camping is especially popular among younger adults, tents are the preferred way to stay on a campsite.[353] Nevertheless, North America is the largest market for camping vehicles, representing 60 % of all newly registered vehicles in 2010. Europe follows with 33 %, and Australasia accounts for 5 %, while South Africa, Japan, China, and others each account for less than 1 %.[354]

These figures show that camping and caravanning is not merely a German peculiarity but a worldwide phenomenon.

In the context of this research and in light of the potential generalization of results, it is important to know whether the sociodemographic characteristics of camping tourists are comparable to those of the general population. The comparison of the age distribution between campers and non-campers shows that there are only very small differences (see Figure 9 below). While the age groups of 30 - 39 and 50 - 59 have a higher share of campers, the age groups above 60 years are less well represented than non-campers. This may be largely due to the fact that a certain level of physical fitness is required to drive to the final destination and to take care of the caravan and oneself on a campsite. Additionally, a camping vacation is usually undertaken with one's spouse.[355] Since the likelihood of widowhood increases with age, the lower share of campers among the elderly is not so much of a surprise. The average age of campers in 2009 was 45.3 years and is slightly below that of non-campers with 46.3 years.[356] The average age of the German population in the same year was even lower with 43.4 years.[357] In comparison, the median age of campers in the US was much lower at 33.0 years.[358] This difference stems partly from the

---

[352] Cf. Outdoor Foundation 2012, pp. 3ff.
[353] Cf. Outdoor Foundation 2012, p. 17.
[354] Cf. Caravaning Industrie Verband e.V. (CIVD) 2012.
[355] Cf. Outdoor Foundation 2012, p. 25.
[356] Cf. Bundesministerium für Wirtschaft und Technologie (BMWi) 2010, p. 18.
[357] Cf. Statistisches Bundesamt Deutschland 2012a, p. 17.
[358] Cf. Outdoor Foundation 2012, p. 13.

lower average age in the US (36.8 years for 2009) but also from the fact that camping in the US is especially attractive to younger people.[359]

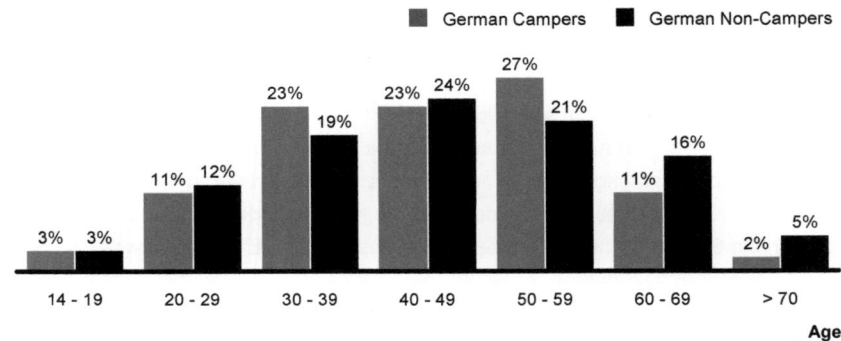

Figure 9: Age Distribution of German Campers versus Non-Campers[360]

Table 4 below compares the net household income and the education level of caravan and motor caravan owners in comparison to the general population in Germany. As one can see caravan owners represent the general population very well, except that the group with the lowest income is underrepresented and the share of caravan owners in the income group of 2,500 - 3,499 EUR is higher than in the general population. The levels of secondary education are comparable. Motor caravan owners differ slightly more from the general population. They tend to have higher incomes (also in comparison to caravan owners) and a higher secondary education level.[361]

---

[359] Cf. United States Census Bureau
[360] Own illustration according to Bundesministerium für Wirtschaft und Technologie (BMWi) 2010, p. 18.
[361] According to Caravaning Industrie Verband e.V. (CIVD) 2013b the average price in 2012 for a motor caravan was 62,617 EUR which is 3.5 times as much as the average price of a caravan at 17,495 EUR. This explains why motor caravans tend to be owned by people with a higher income. For a detailed characterization of camping tourists in the US please refer to Outdoor Foundation 2012.

Table 4: Sociodemographic Characteristics of Caravan Owners Compared to the General Population in Germany[362]

	General Population	Caravan Owner	Motor Caravan Owner
Net household income			
• < 1,500 EUR	30 %	21 %	15 %
• 1,500 - 2,499 EUR	42 %	41 %	45 %
• 2,500 - 3,499 EUR	19 %	27 %	28 %
• ≥ 3,500 EUR	10 %	11 %	12 %
Secondary education			
• Lower secondary education	48 %	52 %	46 %
• Ordinary level	35 %	33 %	32 %
• A level	17 %	15 %	22 %

Despite these small differences, one can state that campers represent the German population fairly well.

5.2 Reasons for Selection of Camping & Caravanning Industry

This research project makes specific demands on the research subject under investigation. For comparisons between older and younger individuals, all age groups should be well represented. As a second requirement, the possibility of user innovations must exist, i.e., the products must not be so complex or technologized that a regular user cannot make modifications anymore.

These two requirements are met by the camping and caravanning industry, as described in the chapters above. All age groups are represented, and the most relevant group for this research project above 50 years is well represented, with 50 - 59 years being the largest group. The broad range of equipment required for camping offers plenty of possibilities for modifications and innovations. Since camping is a leisure activity that people usually undertake during their vacations, they are highly emotionally involved. Additionally, the financial involvement is not to be underestimated, as was shown in chapter 5.1.2. The high financial and emotional involvement is expected to lead to a high level of motivation of individuals to find optimal solutions regarding their needs.[363] If these solutions cannot be purchased, the likelihood of modifying existing products to meet personal requirements is rather high. After all, as shown above, camping is an activity that was invented by users and, therefore, it is a user innovation itself. Lastly, the intensive communications among members of the camping community foster the exchange of information and ideas. The role of the community on user innovations is not part of the research focus

---

[362] According to Caravaning Industrie Verband e.V. (CIVD) 2010.
[363] Ernst et al. 2004, p. 25 also argue, that users of outdoor equipment are generally more deeply involved in their products. This shows in a more thorough product selection process but also in the fact that they tend to repair and modify their products more often.

of this project, but several studies have shown the positive influence of community resources on innovativeness and the quality of the innovations.[364] Camping tourists often have more than one community to interact with. During the warm season, they are at the campsite and come into contact with other campers and some innovations might even be on display to be studied by other campers. Beyond the typical camping season, there are several large online communities in which campers share thoughts, recommendations, and ideas with their community.[365] Altogether, the camping and caravanning market seems to be very suitable for an investigation of user innovators in different age groups.

---

[364] Cf. Franke & Shah 2003; Raasch et al. 2008; Füller et al. 2007; Baldwin et al. 2006, p. 1307.
[365] See chapter 0 for an overview of online camping communities in Germany.

6 Explorative Survey among Companies

6.1 Motivation for Study and Selection of Questions

Before conducting a deeper analysis of user behavior, it is interesting to know whether the phenomenon of user innovation is relevant for managers in the camping and caravanning industry. Additionally, if managers are aware of user innovations in their field, it would be relevant to know whether they have already incorporated ideas from users and whether these collaborations have been successful. In order to do this, a survey among product managers and product development managers of companies that produced camping vehicles and equipment was conducted.[366]

The research project of Dömöter and Franke "Benchmarking of innovation management practices of SMEs in Vienna" from the Vienna University of Economics and Business compared innovation management practices among SMEs in Germany and Austria.[367] Key questions from their project regarding collaboration with users and customers and applied innovation management tools were taken from their questionnaire.[368] Questions about the characteristics of existing user innovations were taken from Lüthje, Herstatt, and Hippel (2005) and Franke, Hippel, and Schreier (2006) and were also included in the user survey (see chapter 7.1.2) to make results comparable.

6.2 Selection of Companies

The bases for the selection of companies were German vehicle manufacturers that are members of the 'Caravaning Industrie Verband e.V.' (CIVD) and German producers of equipment that presented their products at the Caravan Salon 2011 in Düsseldorf. Out of the full list of potential companies, companies with headquarters in Germany were selected. Of this group, dealers were eliminated, so that only manufacturers remained in the list. From the remaining manufacturers, only the ones actually producing equipment for caravans and mobile homes and whose products were evaluated as feasible for user innovations were kept in the sample.[369] Finally

---

[366] Details on the selection of companies are detailed in chapter 6.2.
[367] For more information, please visit
http://bach.wu-wien.ac.at/bachapp/cgi-bin/fides/fides.aspx?search=true;project=true;type=project;tid=1622 (Link valid as of January 22, 2014).
[368] Cf. Dömötör et al. 2007, pp. 45f.
[369] In this step manufacturers of tent equipment were eliminated from the list. Product categories that seemed not to be feasible for user innovators included high technological categories, e.g., GPS tracking devices and satellite receivers.

two double entries were eliminated, so that, in total 85 companies could be contacted (see also Figure 10 below for a detailed waterfall).

All companies received a cover letter and the questionnaire via mail on March 15, 2012. The letters were addressed to the responsible product managers or the product development managers. In case the responsible managers could not be identified, the letter was addressed to the CEO. Replies were made possible via mail or fax. After two weeks, a follow-up reminder was sent via email. Within four weeks, responses from 23 companies were received, resulting in a response rate of 27 %.

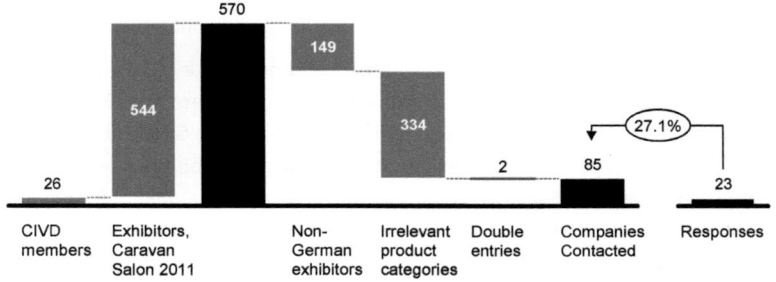

Figure 10: Selection of Approached Camping Companies[370]

6.3 Results of Company Survey

The responses from companies were mostly open-minded and favorable towards user innovation, but they also showed that there is still a lot of room for improvement to integrate customers or users even more.

None of the surveyed camping companies reported to 'always include' customers during the ideation or product development phase (see Figure 11 below). Rather, customers mostly participate during the ideation phase (eleven companies reported 'often'; 20 companies reported 'often' or 'sometimes'). During the later phases of product development, customers are less frequently involved, and only six of the companies reported that they integrate customers often (15 reported 'often' or 'sometimes'). This trend is further reinforced by the fact that only those companies that already integrate their customers often during ideation also do so during product development. Also, all companies, which have commercialized customer ideas before (see Figure 13 below), let their customers participate at least 'sometimes' during the ideation phase. This indicates that some companies are more willing to let

---

[370] Own illustration.

their customers participate in the innovation processes and also to realize these ideas.

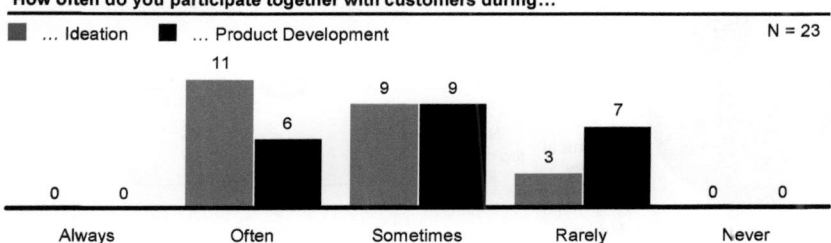

Figure 11: Participation with Customers during Ideation and Product Development[371]

The great majority of companies (83 %) have had customers proactively approached them to propose new product ideas or prototypes. Company representatives were asked to rate the evaluations of these ideas regarding their newness, originality, technical quality, and market potential (see Figure 12 below). The results were rather disappointing because only originality was evaluated positively. On a scale from '1 = very original' to '5 = not original at all', the average for all customer ideas was a 2.9.[372] Newness received an average rating of 3.3, but four out of 19 (equal to 21 %) companies stated that ideas represented generally new products. This share is in line with comparable shares in user innovation studies (see Table 1 above) and should not be underestimated. Technical quality was rated the lowest with an average of 3.9 and no mentions above the neutral statement. This indicates that customers in the camping industry do not focus on technical innovations. Based on the idea that customers often provide ideas for new products to which manufacturers then apply their production expertise to profitably commercialize,[373] there seems to be a lot of room for companies to benefit from customer ideas. Company respondents seem to be more pessimistic about this evaluation, because they rated the market potential low (on average 3.6) and rather non-promising.

---

[371] Own illustration.
[372] The other scales were measured as follows:
Newness: from '1 = totally new product' to '5 = small improvement / modification'
Technical Quality: from '1 = new technology' to '5 = known technology'
Market Potential: from '1 = very high' to '5 = very low'.
[373] Cf. Bogers et al. 2010, p. 868.

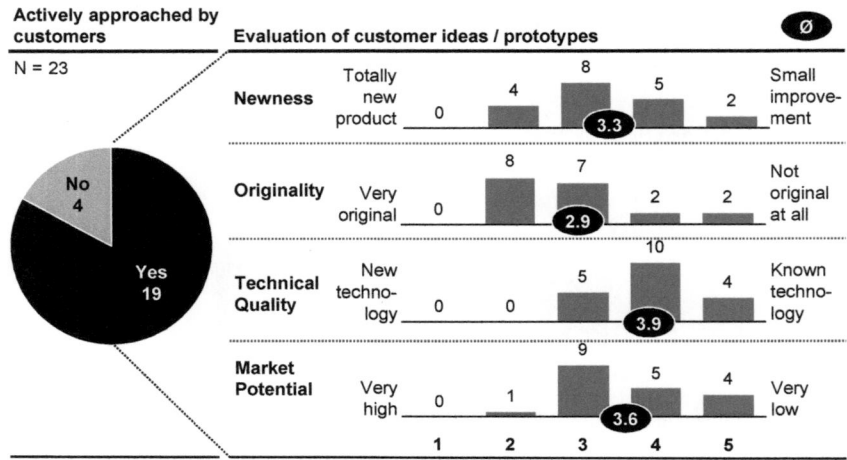

Figure 12: Evaluation of Customer Ideas and Prototypes[374]

17 of the companies stated that they already had realized ideas from customers. The remaining 6 were then asked to state their reasons not to realize existing and known customer ideas (see Figure 13 below). All companies stated that there would be no market demand, which is a questionable statement since at least some customers expressed a need and invested time and effort into creating a solution. 71 % agreed to the statement that a profitable production was not possible. This was followed by the statement that the ideas / prototypes were not technically mature enough (43 %). This relates back to the low score of customer ideas regarding their technical quality above. None of the companies stated that they are generally not interested in customer ideas.

To analyze which tools companies in the field of camping and caravanning apply to integrate customer ideas and opinions into the ideation and product development process, all companies were asked which tools they apply on a regular basis. The most widely prevalent tool is the typical customer survey, which is regularly executed by 18 of the companies. More sophisticated instruments, like a conjoint analysis, the quality-function-deployment or the lead-user method, are applied significantly less often (see Table 5 below). The comparison of applied tools among companies who have already realized customer ideas and those who have not reveals detectable

---

[374] Own illustration.

differences.[375] Companies with realized customer ideas use a simple customer survey less often. Instead, they use conjoint analysis, the lead-user method, and virtual communities much more often than companies that have not yet realized customer ideas.[376] In fact, the lead-user method and virtual user-communities are exclusively used by companies that have already realized customer ideas. From this fact, it can be inferred that companies that apply sophisticated innovation management tools, are also more likely to actually realize customer ideas.

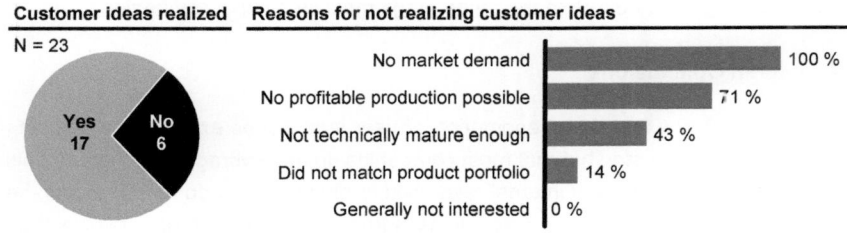

Figure 13: Reasons for Not Realizing Customer Ideas[377]

Table 5: Applied Tools for Customer Integration

Customer integration tool	Overall N = 23	Companies with realized customer ideas N = 17	Companies without realized customer ideas N = 6	Δ
Customer survey	78 %	76 %	83 %	−7 %
Conjoint analysis	22 %	24 %	17 %	+7 %
Quality-Function-Deployment	17 %	18 %	17 %	+1 %
Lead-User method	13 %	18 %	0 %	+18 %
Virtual user-communities	9 %	12 %	0 %	+12 %

In the end, it is crucial for all companies to be competitive and offer attractive products to their customers. One strategy is 'continuous innovation', which necessitates continuously improving upon existing products to always offer innovative products and therefore stay ahead of the competition.[378] An indicator for

---

[375] Inferential statistical methods were not applied, since the rather low sample size did not promise meaningful and robust results.
[376] Späth 2008, p. 26 presented comparable results. According to their research among SMEs 86 % used customer surveys, 12 % quality-function-deployment, 7 % the lead-user method, 6 % conjoint analysis, 5 % virtual user-communities, and 3 % toolkits for user innovations. In their subsample of the most innovative companies these numbers were significantly higher: 96 % used customer surveys, 82 % the lead-user method, 78 % quality-function-deployment, 54 % conjoint analysis, 46 % toolkits for user innovations, and 40 % virtual user-communities.
[377] Own illustration.
[378] Cf. Kotler & Bliemel 2006, p. 688.

the innovativeness of a company is the revenue and profit share of innovations across the past three business years.[379] The 2011 revenue share of innovations in non-research-intensive manufacturing industries was 12 % in Germany.[380] Späth (2008) showed in a survey among SMEs that the most innovative companies have an innovation revenue share of 70 % and a profit share of 77 %.[381] Only one out of the 23 companies in the sample reported a revenue and profit share from innovations between 61 - 80 %. All others reported that their revenue and profit share from innovations is below 20 %. The respondents in the sample are thereby in line with the German average and not within the range of the most innovative companies.

6.4 Interim Conclusions

The analysis shows that the phenomenon of user innovations exists in the camping and caravanning industry, but that most companies do not leverage its potential. This could be caused by a lack of internal resources or because they do not believe in the value of commercializing user innovations. Most companies already involve their customers in the ideation process for new products; some even involve them during product development. Generally, there is still potential for improvement, because none of the responding companies reported to involve their customers by default.

The value of the ideas and prototypes that customers proactively present to companies are evaluated rather negatively, especially with regards to their technical quality and future market potential. A lack of market potential is then also the main reason why customer ideas are not realized. More modern and up-to-date tools (e.g., lead-user method, conjoint analysis) to integrate customers are used only by a minority of companies and are used almost exclusively by companies that have already realized customer innovations.

In summary, one can state that user innovators exist among campers but that their potential for the development of new products is currently underestimated by manufacturers. The following chapter will further analyze what characterizes and motivates user innovators in this field.

---

[379] Cf. Späth 2008, p. 18.
[380] Rammer et al. 2013, p. 9.
[381] Späth 2008, p. 18.

7 Empirical Study among Camping & Caravanning Tourists

7.1 Research Design and Operationalization

The following chapter presents the chosen research strategy for the quantitative study among camping and caravanning tourists. First, structural equation modeling is described as the statistical procedure for data analysis (chapter 7.1.1). Then the theoretical constructs are operationalized and transformed into concrete items[382] (chapter 7.1.2). Finally, the process for the collection (chapter 7.1.3), cleansing, and preparation (chapter 7.1.4) of the data is explained.

7.1.1 Structural Equation Modeling with PLS

Since the late 1970s, the use of structural equation modeling (in the following abbreviated with SEM) has increased in scientific publications of marketing and social sciences.[383] Compared to other multivariate statistical methods[384], SEM allows for highly complex models and the simultaneous analysis of several causal relationships.[385] Additionally, it is easily possible to analyze the relationships between latent variables. Latent variables cannot be observed directly and are therefore measured through manifest (observable) indicators.[386]

An SEM consists of an inner structural model and outer measurement models (see also Figure 14 below).[387] The outer measurement model assigns a value to the latent variables (also called constructs) through analysis of the manifest indicators.[388] The inner structural model represents a set of hypotheses or an entire theory under investigation.[389]

---

[382] The terms *item* and *indicator* are used synonymously, as this is also the norm in the literature on PLS.
[383] Cf. Baumgartner & Homburg 1996, pp. 140f.
[384] E.g., analysis of variance, regression, cluster analysis, conjoint, factor analysis. For an overview, see Hair et al. 2008 or Backhaus et al. 2011.
[385] Cf. Huber et al. 2007, p. 1; Backhaus et al. 2011, p. 517.
[386] Cf. Fornell & Larcker 1981; Huber et al. 2007, p. 1.
[387] Terminology in SEM literature is not always consistent. While CB-SEM literature refers to "structural model" and "measurement model", literature focused on PLS-SEM calls the same "inner model" and "outer model". Cf. Hair et al. 2012, p. 415.
[388] Cf. Fornell & Larcker 1981.
[389] Cf. Hair et al. 2008, pp. 638f.

There exist several forms of SEMs and different approaches, but according to Hair et al. (2008) they all have three characteristics in common:

> "1. Estimation of multiple and interrelated dependence relationships
> 2. An ability to represent unobserved concepts in these relationships and account for measurement error in the estimation process
> 3. Defining a model to explain the entire set of relationships"[390].

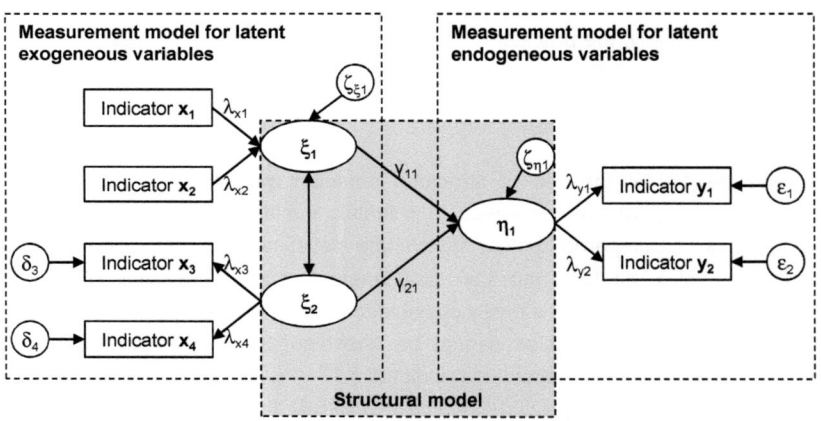

Figure 14: Structural Equation Model with Latent Variables[391]

The two main approaches for the evaluation of SEM are covariance-based SEM (CB-SEM) and variance-based or partial least squares SEM (PLS-SEM).[392] The CB-SEM estimation algorithm optimizes for the fit of the theoretical model with the observed, empirical covariance matrix.[393] The most common algorithms are LISREL (linear structural equations) and AMOS (analysis of moment structures). The PLS-SEM algorithm is more prediction-oriented, explicitly estimating latent variable scores and maximizing for the explained variances of the latent variables.[394] The PLS-SEM approach focuses less on theory testing and more on prediction. As such, it can deal better with highly complex models. In recent years, several publications have dealt with the selection criteria for choosing the correct approach. This is why a detailed

---

[390] Hair et al. 2008, p. 635
[391] Own illustration adapted from Backhaus et al. 2011, p. 519; Jarvis et al. 2003, p. 210; Huber et al. 2007, pp. 5ff.
[392] Cf. Hair et al. 2008, pp. 663f.; Hair et al. 2012, pp. 414ff.; Esposito Vinzi et al. 2010, p. 48; Henseler et al. 2009, pp. 277f.; Tenenhaus et al. 2005
[393] Cf. Schweisfurth 2013, p. 92
[394] Cf. Chin & Newsted 1999.

methodological comparison is not part of this research, and the interested reader should refer to the existing literature.[395]

Chin and Newsted (1999) suggested using the PLS algorithm for research projects that fulfill at least one of the following criteria:

- *"The objective is prediction, and/or*
- *The phenomenon in question is relatively new or changing and the theoretical model or measures are not well formed, and/or*
- *The model is relatively complex with large numbers of indicators and/or LVs, and/or*
- *The data conditions relating to normal distribution, independence, and/or sample size are not met."*[396]

For the following reasons, the PLS approach to SEM was applied:

1) Focus on prediction, not on theory testing

If the research goal of a project is on theory testing or confirmation, CB-SEM should be used.[397] If the focus is on predicting a dependent construct and the key drivers or if the research is exploratory, PLS-SEM should be applied.[398]

Research on user innovations, especially in the area of consumer goods, has been conducted since the 1990s and is therefore still a rather young research area.[399] The explicit analysis of whether user innovators exist within certain age groups and whether there are differences in the determinants of innovative behavior regarding these age groups has not been the focus of research before. Using PLS-SEM is recommended in such an environment because it "*[...] maximizes the explained variance of all dependent variables [...]*"[400] and therefore increases the model fit with the data. In comparison with CB-SEM, the outer measurement model is often overestimated (i.e., the weights between indicators and latent variables are higher than in reality), while the inner structural model is underestimated. This phenomenon is usually referred to as "PLS bias".[401] Simulation studies have shown that this bias is

---

[395] For a comparison of both approaches, see Fornell & Bookstein 1982; Chin & Newsted 1999; Harhoff et al. 2003, pp. 629ff.; Bliemel et al. 2005; Henseler et al. 2012b; Hair et al. 2011; Hair et al. 2012.
[396] Chin & Newsted 1999, p. 337. Similar catalogues to decide between CB-SEM and PLS-SEM can be found in Hair et al. 2011, p. 144; Henseler et al. 2012b, pp. 262f.
[397] Henseler et al. 2009, pp. 296f.
[398] Hair et al. 2011, p. 144.
[399] See also chapter 3.
[400] Henseler et al. 2009, p. 297.
[401] Cf. Henseler et al. 2012b, p. 263.

negligible in practical applications and the effect diminishes with larger sample sizes.[402] Nevertheless, PLS-SEM estimates usually show a lower degree of variance than CB-SEM, which prompts more robust results (especially regarding cases in which assumptions are violated).[403]

2) Suitable for complex models

The PLS algorithm requires less computational resources and is therefore able to handle more complex models with many latent variables, indicators, and relationships.[404] The presented model consists of six latent variables, 15 indicators, and more than 15 relationships in the structural model.

3) No distributional assumptions required

Most multivariate analysis methods often require data that is normally distributed.[405] Often, this requirement cannot be met, and non-parametric approaches need to be applied. Chapter 0 also shows that some of the data for this research is not normally distributed, which supports the use of PLS.

4) Small sample sizes possible

Many researchers point out that, although PLS is able to calculate results from small sample sizes, it is not a universal remedy for very small samples, and conclusions regarding generalization must be drawn consciously and carefully.[406] Chin (1998b) introduced the 'rule of ten' as the minimum criteria for the sample size.[407] This rule states that the required sample size should be at least ten times the maximal number of the path coefficients, pointing towards a latent variable. The research model of this study (see Figure 27 below) contains five path coefficients pointing toward 'innovative behavior'. According to the 'rule of ten', a sample size of at least 50 is required. The actual sample size is 351, which is well above the required minimum. By splitting the sample for multi-group analysis (e.g., based on age cohorts), the overall sample size might drop significantly, so that the advantages of PLS are still valid.

---

[402] Cf. Hair et al. 2011, p. 143; Reinartz et al. 2009, p. 338.
[403] Cf. Henseler et al. 2012b, pp. 263f.; Reinartz et al. 2009, p. 340.
[404] Cf. Chin & Newsted 1999, p. 335.
[405] Cf. Hair et al. 2008, p. 71.
[406] Cf. Marcoulides & Saunders 2006, p. iii; Hair et al. 2011.
[407] Cf. Chin 1998b.

5) Use of formative and reflective measures[408]

Contrary to CB-SEM, PLS can also handle formative measures.[409] Out of the six latent variables, 'use experience' is operationalized as a formative construct based on two indicators (see also chapter 7.1.2.1).

Based on the stated reasons above, the variance-based approach and the PLS algorithm were applied. All analyses were conducted with SmartPLS[410] and PASW Statistics 18[411].

7.1.2 Operationalization of Constructs

All constructs and indicators in this research project have been applied in previous studies and have been proven to create reliable and valid results. An overview of the main constructs and underlying indicators used in the survey is provided in Table 6 below. Most indicators could be used in their original form, but some had to be adjusted so that they are specific to the camping and caravanning environment.

Since all questions in the survey had to be in German, all constructs that were originally in English were translated into German by the author and translated back to English by two native speakers. Results were compared and adjustments to the wordings were made where necessary.

A pretest with experts (N = 2) and non-experts (N = 3) in the subject was conducted to evaluate the overall structure of the survey, comprehensibility of questions, and the required time for the survey. Small wording adjustments to a few questions were made to improve understandability of the survey, but none of the indicators for the key constructs had to be changed.

---

[408] The direction of causality defines the difference between formative and reflective measurements. While reflective indicators are the consequences of the underlying construct they are trying to measure, formative indicators are the underlying cause for the construct. Cf. Rossiter 2002, p. 314; Diamantopoulos & Winklhofer 2001, p. 269; Jarvis et al. 2003, p. 203.
[409] Cf. Diamantopoulos & Winklhofer 2001, p. 274.
[410] Version 2.0.M3. Cf. Ringle et al. 2005.
[411] Version 18.0.0 (30.07.2009).

Table 6: Operationalization of Constructs[412]

Code	Item
Use Experience	
UE [1]	How many days per year do you do camping?
UE [2]	For how many years have you been camping?
Product Knowledge	
PK [1]	I use my equipment intensely.
PK [2]	I have a good overview of the available equipment on the market.
PK [3]	I am well versed in the materials of my equipment.
Technical Expertise	
TE [1]	I can repair my own equipment.
TE [2]	I can help other campers solve problems with their equipment.
TE [3]	I am handy and enjoy tinkering.
TE [4]	I can make technical changes to my camping equipment on my own.
TE [5][†]	I always try to keep up to date with my equipment with regard to the materials, innovations, and possibilities.
TE [6][†]	I am a huge fan of the technical aspects of this area.
TE [7][†]	I come from a technical background in my profession or education (e.g., engineering).
Lead Userness	
LU [1]	I usually find out about new camping products and solutions earlier than others.
LU [2]	I have benefited significantly by the early adoption and use of new camping products.
LU [3][†]	I have tested prototype versions of new camping products for manufacturers.
LU [4]	Among campers, I am regarded as being on the "cutting edge".
LU [5]	I have new needs which are not satisfied by existing camping products.
LU [6]	I am dissatisfied with the existing camping equipment.
Innovative Behavior	
IB [1]	Have you improved existing products or had ideas for new products that were not offered on the market before?
IB [2]	How far have you developed your idea to date?
Chronological and Cognitive Age	
Chronological Age	How old are you?
FEEL Age	I FEEL as though I am in my ...
LOOK Age	I LOOK as though I am in my ...
DO Age	I DO as though I am in my ...
INTEREST Age	My INTERESTS are mostly those of a person in his/her ...

[†] Omitted after confirmatory factor analysis (see chapter 7.2.6.1)

7.1.2.1 Main constructs

Use Experience

Use experience was measured formatively with two items, as is the common approach.[413] The first item is the frequency of use, measured in days per year. The second item is the total duration of use experience measured in years. Both items were measured without a pre-defined scale, but with an open text field.

---

[412] Only constructs for the evaluation of the structural model are shown here. The full survey, including additional items regarding motivational factors, innovation characteristics, and demographics, can be found in Appendix 10.

[413] Cf. Lüthje 2004; Lüthje et al. 2005; Schweisfurth 2013; Schreier & Prügl 2008; Franke & Shah 2003 labeled it "time in community"; Slaughter 1993 focused solely on the duration indicator.

Product-related Knowledge

"*Product related knowledge consists of know-how about the product architecture and the used materials and technologies of the existing products in the market.*"[414] This knowledge is required to translate a user's often implicit needs and demands into explicit detailed (technical) specifications of requirements.

In research into online communities, product-related knowledge has often been measured directly by counting the number of technical terms used or by semantic analysis of a user's posts.[415] Since users should be questioned directly, this was not feasible. Therefore, the approach of Lüthje (2000) was followed, who measured product-related knowledge reflectively with three items on a 5-point Likert scale (from "strongly agree" to "strongly disagree").[416]

Technical Expertise

The full seven item construct established by Franke, Hippel, and Schreier (2006), measured on a 5-point Likert scale from "strongly agree" to "strongly disagree", was applied.[417] Franke, Hippel, and Schreier (2006) eliminated three items after tests for validity. All seven original items were initially applied in this study, but after testing for reliability and validity of the construct (see chapter 7.2.6.1), the shorter four-item was used during the analyses.

Lead Userness

According to literature, lead users show two characteristics: "being ahead of trend" and "high benefits" (see chapter 3.4 for more details).[418] Morrison, Roberts, and Midgley (2004) showed that both characteristics are continuously distributed within their sample of innovators and non-innovators, and, therefore, it should not be measured dichotomously.[419] They also argued that both factors are significantly correlated and "*[...] do form part of the same construct*"[420]. Franke, Hippel, and Schreier (2006) have challenged this and have argued that both characteristics are actually independent dimensions that should be measured formatively.[421]

---

[414] Lüthje 2004, p. 686.
[415] Cf. Marchi et al. 2011; Füller et al. 2007.
[416] Cf. Lüthje 2004 also.
[417] Cf. Schweisfurth 2013 also.
[418] Cf. Hippel 1986, p. 796.
[419] Cf. Morrison et al. 2004, p. 358, 1999, p. 24.
[420] Morrison et al. 2004, p. 375.
[421] Cf. Franke et al. 2006, pp. 303f.

Nevertheless, most researchers used a combined construct for lead userness and measured all items reflectively, fulfilling all common quality criteria.[422]

The "being ahead of trend" characteristic is often measured with the concrete actions or achievements of the respondent.[423] While this procedure is appropriate for sporting activities, it is not very feasible for leisure activities like camping. Therefore, the lead userness questionnaire developed by Franke and Shah (2003) was applied. It consists of five items for "being ahead of trend" and two items for "high expected benefits".[424] One item from the "being ahead of trend" battery of questions had to be dropped, because discussions with experts showed that it was not suitable to the realm of camping and caravanning.[425] The items were measured on a seven-point Likert scale from "strongly agree" to "strongly disagree".[426]

Innovative Behavior

Innovative behavior was measured with a single item (IB [1]), which could either be answered with "yes" or "no".[427] To assure, that all respondents had a common understanding of what is considered an innovation, the question was followed by a short explanation.[428] Using a single item measurement is always problematic and should only be used if the construct to be measured is very concrete and low in complexity.[429] Both are true in the case of innovative behavior.

All respondents who indicated that they had already shown innovative behavior received a follow-up question asking them to further detail the highest development stage their innovation was in (IB [2]). For this item, the scale of Lüthje, Herstatt, and Hippel (2005) was expanded to include an answer option for already commercialized innovation. The final scale comprised the following five steps: (1) I have a possible solution in mind; (2) I have made concept descriptions/drawings; (3) I have built a prototype that is reliable enough for me to use it; (4) Others are using prototypes

---

[422] Cf. Schweisfurth 2013; Morrison et al. 2004; Schreier & Prügl 2008; Schreier et al. 2007; Kratzer & Lettl 2008; Lüthje 2004 for reflective measurement. Jeppesen & Frederiksen 2006 and Jeppesen & Laursen 2009 measured lead userness formatively, but did not distinguish between the two characteristics.

[423] Cf. Schweisfurth 2013, p. 129; Franke et al. 2006, pp. 306f.; Schreier & Prügl 2008, p. 340.

[424] Cf. Franke & Shah 2003, p. 163.

[425] The last item "I improved and developed new techniques in ..." was dropped.

[426] Franke & Shah 2003 used a seven-point Likert scale from "very accurate" to "not accurate at all". The scale was adjusted to match it to the wording of the Likert scales in the questionnaire to limit the potential confusion of respondents and reduce measurement error.

[427] Compare for example Urban & Hippel 1988; Franke & Shah 2003, p. 176; Lüthje 2000.

[428] The explanation read: "A product idea/improvement could be linked to an already existing product or a radical new development." (German original: "Eine Produktidee/-verbesserung kann sich auf ein bereits bestehendes Produkt beziehen oder eine völlige Neuentwicklung sein.")

[429] Cf. Fuchs & Diamantopoulos 2009, p. 203; Weiber & Mühlhaus 2010, p. 92.

based on my idea; (5) The idea was already commercialized and is available on the market.

Chronological and Cognitive Age

Chronological age is simply measured by asking the respondent to state his or her age in years. If the time of the innovation dates back several years, the current chronological age does not represent the actual age during the innovation, and the allocation of innovators to the specific age cohorts will be flawed. In order to correct for this potential bias, innovators were asked to state the year of their last innovation.[430]

Cognitive age is measured by the four dimensions (*FEEL* age, *LOOK* age, *DO* age, and *INTEREST* age) developed by Barak and Schiffman (1981), which have since been proven to be valid and reliable.[431] It was shown that all four dimensions of cognitive age measure the same underlying construct.[432] The LOOK age component of the cognitive age construct has shown a lower reliability, and dropping this part has been suggested, since it does not significantly influence overall fit and reduces the effort needed for data collection and analysis.[433] Since this was not a concern for this research project, the LOOK age component was kept in the survey.

It can generally be assumed that cognitive age is a universally applicable construct that creates consistent and comparable results across different cultures and countries.[434] To make sure that respondents understood the questions correctly and were prepared to think about the different dimensions of cognitive age, the questions were preceded by an introductory explanation.[435]

Several different scales have been used to measure cognitive age: semantic differential, ratio, and Likert scales. Auken and Barry (1995) and Auken, Barry, and Bagozzi (2006) compared these three scales with regard to their trait, error, and method variance. They concluded that the semantic differential scale provides results with the highest trait validity. They recommended using it in future research, but they

---

[430] Asking for the last innovation is most appropriate because the research project wants to analyze whether people are still innovating at a higher age. Cf. Morrison et al. 2000, p. 1515.
[431] Cf. Barak & Schiffman 1981; Henderson et al. 1995; Wilkes 1992; Auken & Barry 1995.
[432] Cf. Auken & Barry 1995, p. 114; Auken et al. 2006, p. 445.
[433] Cf. Wilkes 1992, p. 298.
[434] Barak 2009 provides an overview of studies across 18 cultural disparate countries. In all of them cognitive age showed reliable and valid results and cognitive age was always lower than chronological age.
[435] The introductory explanation according to read: "*Most people seem to have other 'ages' besides their official or 'date of birth' age. The questions which follow have been developed to find out about your 'unofficial' age. Please specify which age group you FEEL you really belong to:*" Barak & Schiffman 1981, p. 605.

ignored the fact that only a ratio scale provides results that can be compared to the true chronological age of a person. Therefore, a variant of the ratio scale was applied: a half-decade scale, which is based on the original full-decade scale but provides more detailed results.[436] This type of scale also proved to be more reliable and robust across different countries.[437] The scale consisted of 14 steps, "20 - 24 years" being the lowest and "85 - 89" being the highest. In order to transform these discrete values into interval data, the mid-point for each half decade was inserted for each respondent and cognitive age dimension (e.g., 87 for "85 - 89 years").[438] The cognitive age score was then computed as an unweighted average of the four sub-dimensions.[439] With this average and the chronological age, the difference could be computed by subtracting the chronological age from the cognitive age. A negative magnitude connotes that the respective person feels younger than they actually are.[440]

7.1.2.2 Innovation Characteristics

A further target of this research study was to analyze whether the innovations of older and younger people differed in their characteristics. In order to do that, items were included that measured attractiveness and other qualities of innovations. Innovators were asked to think of their last innovation and refer to it when providing answers for the following questions. According to Lüthje, Herstatt, and Hippel (2005), one can rely on user innovators' self-evaluations because although innovators *"evaluate the commercial potential of their innovations slightly more positively than [...] independent experts, the level of difference was not statistically significant."*[441] Innovation attractiveness was measured through four items by Franke, Hippel, and Schreier (2006), regarding the benefit of the innovation to other campers today and in the future (measured on a five-point Likert scale from "very high" to "very low") and the sales potential today and in the future (measured on a five-point Likert scale from "many" to "a few").[442] Additionally three qualities of the innovations were inquired about from the innovators: newness, technical quality, and creativity. Measures for newness and technical quality were taken from Lüthje, Herstatt, and Hippel (2005), only changing the scale from a seven-point to a five-point Likert scale, so that it

---

[436] A full-decade scale was used by Barak & Schiffman 1981; Wilkes 1992; Henderson et al. 1995, the half-decade scale was introduced by Cleaver & Muller 2002.
[437] Cf. Barak et al. 2011, p. 480.
[438] Cf. Cleaver & Muller 2002, p. 231.
[439] Cf. Barak & Schiffman 1981, p. 604.
[440] Cf. Cleaver & Muller 2002, p. 231.
[441] Lüthje et al. 2005, p. 958.
[442] Cf. Franke et al. 2006, p. 310.

matched the rest of the survey.[443] The item for creativity was introduced by Franke, Hippel, and Schreier (2006). It measures creativity on a five-point Likert scale from "very creative" to "not creative at all".[444] Finally innovators were asked how they would classify their innovation (answer options: comfort improvement, interface improvement, cost savings, new functionality, time improvement, others).[445]

To obtain information about the innovation process, respondents were confronted with questions regarding the required development time (ratio scale: < 1 week, 1 - 2 weeks, 2 weeks - 1 month, 1 - 3 months, 3 - 6 months, 6 - 12 months, > 12 months), the development frequency (ordinal scale: almost all my equipment, most of the time, sometimes, rarely, only this time), cooperation during the development phase (ordinal scale: alone, collaborative – I was the driving force, collaborative – All participated equally, collaborative – Someone else was the driving force), and cooperation during realization (dummy coded: yes = 1, no = 0). Lastly, respondents were asked to briefly describe their product idea in a free text field.[446]

7.1.2.3 Motivational Factors

Motivation, especially intrinsic motivation, has been identified as one of the key drivers of innovative activity of users.[447] It has been shown that intrinsic motivators are more important than extrinsic motivators.[448] The impact of motivational factors on innovative behavior and the resulting innovations was not the focus of this study. Rather, the focus was on whether there exist relevant differences between age groups. Therefore, out of the multitude of available measurement scales[449] only basic questions that had been successfully applied in lead user research before were applied.

Respondents who had reported an innovation were asked to indicate their agreement to two statements regarding extrinsic motivation ("I wanted to earn money with the idea." and "I was paid well for my assistance.") and three statements regarding intrinsic motivation ("I wanted to use the product myself", "It was nice to receive

---

[443] Cf. Lüthje et al. 2005, p. 957. Newness was measured on a scale from "totally new product" to "small improvement / modification". Technical quality was measured on a scale from "New technology / High-tech solution" to "Known technology / Low-tech solution".
[444] Cf. Franke et al. 2006, p. 310.
[445] Response options were developed based on expert interviews and feedback from the pretest.
[446] Please refer to Appendix 10 for the complete survey questionnaire.
[447] Cf. Franke & Shah 2003, p. 158; Hienerth 2006, p. 286; Marchi et al. 2011, p. 351; Baldwin et al. 2006, p. 1296; Lettl et al. 2006, pp. 32f.; Hippel 2005a, pp. 60f.
[448] See for example Franke & Shah 2003, pp. 173f.. Lüthje 2000, p. 69 even found a negative effect of extrinsic motivation on innovative behavior.
[449] A meta-analysis by Mayer et al. 2007 identified approximately 230 scales related to motivation alone in the database PsycINFO® by the American Psychological Association.

recognition.", and "It was fun to improve my equipment.").[450] Responses were measured on a five-point Likert scale from "strongly agree" to "strongly disagree".

7.1.2.4 Control Variables

Variables that are neither dependent nor independent variables of a research model, but still need to be analyzed in order to fully understand the interdependencies within a research model, are called covariates or control variables.[451]

Demographical data was taken from the Mikrozensus 2011 whenever applicable.[452] This included the questions for gender (dummy coded: male = 1, female = 0), marital status (nominal scale: single, in a partnership, married, divorced, widowed); monthly net household income (ratio scale: < 1,000 EUR, 1,000 - 2,000 EUR, 2,000 - 3,000 EUR, 3,000 - 4,000 EUR, 4,000 - 5,000 EUR, > 5,000 EUR)[453], highest academic degree / vocational qualification (ordinal scale: Secondary general school certificate, intermediate school certificate, entrance qualification for universities or universities of applied sciences, apprenticeship, degree of a university, doctor's degree), and current or last job (open question plus nominal scale: salaried employee, wage earner, apprentice, self-employed, family worker, public official / judge, soldier, person doing a side job). One additional question regarding the occupational category was included based on ISCO 08 (nominal scale: 0-armed forces occupations, 1-managers, 2-professionals, 3-technicians and associate professionals, 4-clerical support workers, 5-service and sales workers, 6-skilled agricultural, forestry, and fishery workers, 7-craft and related trades workers, 8-plant and machine operators, and assemblers, 9-elementary occupations).[454]

Since chronological age by itself does not indicate much about lifestyle and significant events in one's life[455], additional control variables were added to the questions, which have not been used in previous studies.

The question whether someone is still employed or already retired is regulated in most countries and therefore correlates strongly with chronological age.[456] In the social sciences there exist several very detailed scales which are also used in

---

[450] Cf. Lüthje 2000; Franke & Shah 2003, p. 177.
[451] Cf. Bortz & Schuster 2010, pp. 26f.
[452] Cf. Statistische Ämter des Bundes und der Länder. Some answer options were combined for reasons of more simplicity and with regard to the length of the survey.
[453] The midpoint of ranges was used for statistical analysis, e.g., for the "1,000 – 2,000 EUR" range, 1,500 was used. 500 was used for the "< 1,000 EUR" option, 5,500 for the "> 5,000 EUR" option.
[454] Cf. Statistik Austria 2011.
[455] See chapter 2.2 - 2.6 for further details.
[456] Cf. Backes & Clemens 2008, pp. 60ff.. Disney & Johnson 2001, p. 11 show in an overview of the most important OECD countries, that the normal pension age lies between 60 and 65.

censuses, but these scales are too detailed and lengthy for this research project.[457] Therefore, a simplified ordinal answering scale for occupational status, which contained four answer options (full time, part time, unemployed, retired) was used.

Since older people usually do not have to cater for their children anymore, and they have already made their largest investments, the share of net income they can freely dispose of should be higher than for younger people.[458] Therefore, a "disposable income" variable was introduced, asking respondents to state the percentage of household income they can freely dispose of. The metrical variable was measured on a ratio scale from 0 to 100 %. Along with more flexibility in spending, older people are also thought to have more free time, since they might not work full-time anymore, and do not need to care for their children anymore.[459] To measure this, the metric variable "disposable time" was introduced by asking respondents to state the average amount of hours they can freely dispose of during a regular day between 8 a.m. and 11 p.m. (ratio scale from 0 to 15).

A very Germany-specific issue is the division of the nation that occurred between 1949 and 1990. Especially older people who grew up in the two separate countries experienced a different socialization and were confronted with a growing disparity in the availability of products and services.[460] This difference could have had an influence on their general attitude towards innovative behavior because people in the former GDR more often had to face issues of economic scarcity. To test for this fact, a nominal variable "origin" was introduced, composed of the answer options "Area of the former GDR", "Western Germany", and "Others".

7.1.3 Data Collection and Sample Description

There exist qualitative and quantitative research strategies.[461] Quantitative research follows a deductive approach and is usually used to test theories.[462] Therefore,

---

[457] See for example questions 30 – 39 in the German Mikrozensus 2011, cf. Statistische Ämter des Bundes und der Länder 2011.
[458] See for example the life cycle hypothesis of savings by Ando & Modigliani 1963 and the life cycle consumption model of Skinner 1988. Both argue that older individuals consume a higher share of their income compared to younger individuals because individuals tend to balance their consumption over time to maximize their utility. The savings rate is therefore higher during the middle third of one's life cycle. Additionally the future income of older individuals is more certain so their need for precautionary savings are lower.
[459] Retired people in Germany spend on average 5 hours and 30 minutes less per day on employment and training compared to fully-employed people. They invest large parts of this time in hobbies and sports (+ 1 hour and 40 minutes), unpaid work (+ 2 hours and 10 minutes), and cultivating contacts, entertainment (+ 18 minutes). Cf. Statistisches Bundesamt Deutschland 2013.
[460] Cf. Frese et al. 1996, p. 55 and see Alesina & Fuchs-Schündeln 2007.
[461] Cf. Bryman 2008, pp. 21ff.; Bortz & Döring 2009.
[462] Cf. Bryman 2008, p. 22.

quantitative methods to collect data and test the hypotheses were chosen. It is estimated that around 90 % of all data in the social sciences is collected via surveys.[463] This approach was also chosen for this research and a self-completion questionnaire was used for the main part of this study. Table 7 shows the advantages and disadvantages of self-completion questionnaires over structured interviews. Although the disadvantages outnumber the advantages, the performance of self-completion questionnaires regarding time, cost, and reachability of respondents outweighs the disadvantages. Additionally, certain precautionary measures were applied to mitigate the disadvantages. Simple and clear terms and questions were used, complex and hypothetical questions were avoided, key terms were explicitly defined, and free text fields for additional comments were included.[464] Additionally, the online survey was split into parts that could not be skipped or returned to, and respondents were only able to proceed to the next section if all mandatory questions were answered.

Table 7: Evaluation of the Self-Completion Questionnaire in Relation to the Structured Interview[465]

Advantages	Disadvantages
• Cheaper to administer • Quicker to administer • Absence of interviewer effects • No interviewer variability • Convenience for respondents • Greater geographical reach	• Cannot prompt • Cannot probe • Cannot ask many questions that are not salient to respondents • Difficulty of asking other kinds of questions • Questionnaire can be read as a whole • Do not know who answers • Cannot collect additional data • Difficult to ask a lot of questions • Not appropriate for some kinds of respondent • Greater risk of missing data • Lower response rates

Many camping and caravanning tourists inform themselves online before upcoming travels and there exist several online communities that foster the exchange of knowledge among campers. To facilitate the data collection process and benefit from the low cost of online surveys, data was collected from members of these communities. To adjust for the lower number of older people, who participate in online communities, a paper-based survey with the exact same questions was conducted on campsites. The following two chapters describe the selection of communities and campsites as well as the process of data collection.

---

[463] Cf. Bortz & Döring 2009, p. 236.
[464] Cf. Porst 2000, p. 2.
[465] According to Bryman 2008, pp. 217f.

7.1.3.1 Online Survey

An extensive Google search was conducted to identify the most relevant German-speaking online camping communities.[466] Figure 15 provides an overview of the twelve largest and most important communities, sorted according to the activity of their members, based on average posts per member.

All community administrators were contacted via email or a contact form, asking them whether they would support the research project either by promoting the link to the survey themselves or by allowing to have the linke posted to the forum. Six out of the twelve community administrators responded positively, and members of the following communities were asked to provide input to the survey: Campen.de, Camperfreunde.com, Wohnwagen-forum.de, Camperboard.de, ClassiCaravan, and Klappcaravanforum.de.[467]

Community-specific survey links were posted from May 10th to 30th, 2012, and the surveys were open for approximately three weeks for each community. The last survey was closed on June 29th, 2012.

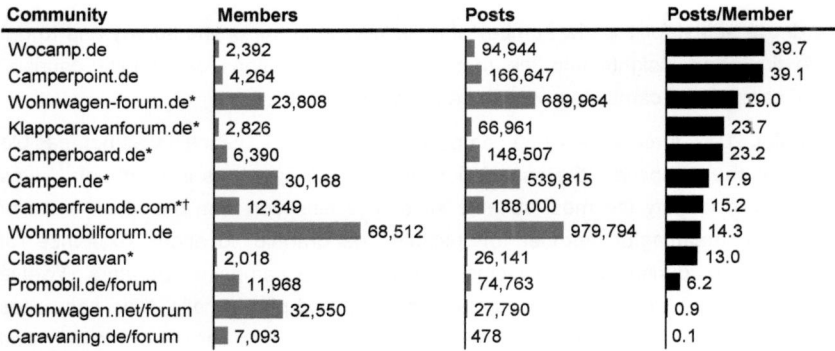

Community	Members	Posts	Posts/Member
Wocamp.de	2,392	94,944	39.7
Camperpoint.de	4,264	166,647	39.1
Wohnwagen-forum.de*	23,808	689,964	29.0
Klappcaravanforum.de*	2,826	66,961	23.7
Camperboard.de*	6,390	148,507	23.2
Campen.de*	30,168	539,815	17.9
Camperfreunde.com*†	12,349	188,000	15.2
Wohnmobilforum.de	68,512	979,794	14.3
ClassiCaravan*	2,018	26,141	13.0
Promobil.de/forum	11,968	74,763	6.2
Wohnwagen.net/forum	32,550	27,790	0.9
Caravaning.de/forum	7,093	478	0.1

* Final Participants
† As of August 10, 2011. Community was relaunched since 2011 and old members and posts were deleted.

Figure 15: Online Camping Communities in Germany (as of February 24th, 2014)[468]

---

[466] Key search terms comprised of "Camping", "Caravaning", "Wohnmobil", "Wohnwagen", "Forum", and "Community".
[467] Administrators of Promobil.de/forum, Wohnwagen.net/forum, and Caravaning.de/forum never responded to the inquiry. Administrators of Wocamp.de, Camperpoint.de, and Wohnmobilforum.de did not want to support the research project.
[468] Own illustration.

Table 8: Responses from Online Survey

Community	Views	Clicks	Click-Rate	Complete Surveys	Completion Rate
Campen.de	427	48	11.2 %	28	6.6 %
Camperfreunde.com	n/a	23	n/a	12	n/a
Camperboard.de	28	9	32.1 %	3	10.7 %
ClassiCaravan	65	18	27.7 %	10	15.4 %
Klappcaravanforum.de	1,873	175	9.3 %	111	5.9 %
Wohnwagen-forum.de	446	141	31.6 %	91	20.4 %
Total[a]	2,839	391	13.8 %	243	8.6 %

a) Excluding Camperfreunde.com

The response rates (based on completed surveys and thread views) ranged from 5.9 % to 20.4 % and the overall response rate was 8.6 %. This response rate is satisfying and within expectations for an online survey conducted without directly addressing the respondents and without incentivation.[469]

Online survey respondents' ages ranged from 19 to 79 and averaged 45.9 years. The share of respondents of at least 55 years was 15.6 %. More descriptive information on the sample can be found in chapter 7.2.1.

7.1.3.2 Paper-based Survey

In order to adjust for the low number of older people in the online survey and to get some first-hand insights into the motivations and situations of camping tourists, actual campers on campsites were to be surveyed directly.

As of May 2012, there were 2,859 campsites in Germany.[470] Since it was not feasible to visit all the campsites, the association of campsite operators in Germany[471] was consulted to identify the most suitable sites. The selection was based on a set of criteria: a meaningful number of pitches, geographic location, existence of representative tourists, and expected support by the campsite operators. Twelve campsites were selected and contacted via email and phone calls. Nine campsites agreed to support the research project, and the survey was carried out in July and August 2012 (see Figure 16 below).

---

[469] Cf. Shih & Fan 2008, pp. 259 & 265.
[470] Statistisches Bundesamt Deutschland 2012b, p. 21.
[471] Bundesverband der Campingwirtschaft in Deutschland e.V. (BVCD).

Research Design and Operationalization 91

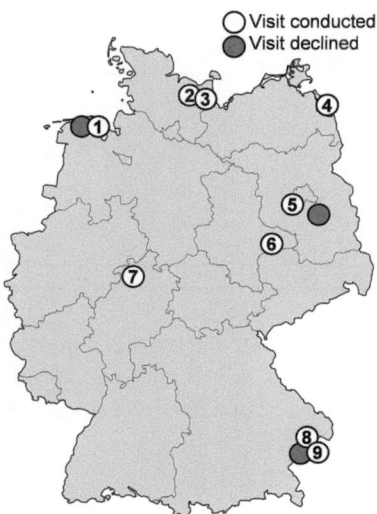

Figure 16: Overview of Approached Campsites in Germany[472]

Distribution and collection of the surveys always followed the same procedure. Every campsite had a central shop or bakery which usually opened between 7 and 8 a.m. The author of this dissertation positioned himself in front of the shop and approached all people walking by and introduced them shortly to the research project. They were then asked whether they would be willing to fill out a survey. In case they agreed, a survey was handed to them, which they did not have to fill it out right away but could do so in their own camper. By doing so, potential bias due to a present interviewer was limited, and the situation was more similar to the online survey, where respondents also were not able to ask follow-up questions or make oral statements, which would not be captured on the response sheet.[473]

Table 9 below provides an overview of the distributed surveys and responses for each campsite. The total number of people approached was not tracked but is approximately two or three times greater than the distributed surveys. Differences in the amount of distributed surveys, as well as the response rate, can be explained with the different weather conditions. During the appointment in Herzhausen (Vöhl) and Schlaitz, for example, the weather was cold, and there was much rain, so most people did not leave their campers and rather stayed inside.

---

[472] Own illustration.
[473] Cf. Bortz & Döring 2009, pp. 246ff.

Table 9: Responses of Paper-based Survey

Nr.	Campsite	Distributed Surveys	Responses	Response Rate
1	Schillig	33	20	60.6 %
2	Plön	23	19	82.6 %
3	Eutin	14	10	71.4 %
4	Karlshagen	18	11	61.1 %
5	Potsdam	22	14	63.6 %
6	Schlaitz	11	6	54.5 %
7	Herzhausen (Vöhl)	8	1	12.5 %
8	Bad Füssing	38	28	73.7 %
9	Bad Griesbach	68	30	44.1 %
	Total	235	139	59.1 %

Paper-based survey respondents' ages ranged from 19 to 86 and averaged at 61.0 years. The share of respondents of at least 55 years was 79.4 % and was, hence, much higher than in the online sample. More descriptive information on the sample can be found in chapter 7.2.1.

7.1.4 Data Cleansing and Preparation

7.1.4.1 Missing Data

There exist three reasons (usually called missingness mechanisms) in the literature that account for why data points can be missing in a data set: Missing Completely at Random (MCAR), Missing at Random (MAR), and Missing Not at Random (MNAR).[474] The missingness is MCAR if the occurrence of missingness is not related to any of the observed or unobserved variables.[475] If the missingness mechanism is MAR, the occurrence of missingness can be fully explained by the remaining variables.[476] MAR missingness occurred when respondents would not disclose their level of income, but it could be approximated by their job type and level of education. Missing data on the details of the innovations are also MAR, if the respondent has indicated that he or she did not innovate at all, which would account for the missing details. MNAR is present if the probability of the occurrence of missingness depends on the actual value of the variable itself and no other variables.[477] In the underlying data set, MNAR missingness occurred when respondents would not disclose a detailed description of their innovation because they worried about their IP rights. In this case, the evaluation of the innovation and its characteristics could not be verified.

---

[474] Cf. Cole 2008, pp. 216f.
[475] Cf. Abraham & Russell 2004, pp. 315f.
[476] Hair et al. 2008, p. 49.
[477] Cf. Cole 2008, p. 217.

Underlying causes for missing data can be manifold and need to be considered while discussing potential reasons for missing data and deciding upon consequences. Potential causes for missing data could be the survey itself (e.g., unclear questions, inappropriate answer options, too lengthy), the respondent (e.g., distraction, shame, excessive demands on respondent), but also technical influences, like data loss in IT systems and incomplete data entry.

The initial data set contained 553 cases (139 from the paper-based survey and 414 from the online survey) and 60 variables. In the online survey, there were 148 respondents who only clicked on the survey link or looked at the first page without leaving any answers.[478] These 148 cases were immediately and without further analysis deleted from the sample. In the remaining sample of 405 cases, the range of missing values within the variables ranges from 0 % to 19 %. Comparing the individual cases, the share of missing values ranges from 0 % to 54 %.

To be able to conduct meaningful analyses, casewise deletion was conducted for all cases with a share of missing values of 20 % or higher.[479] Finally, 17 cases from the paper-based survey were removed because lead user characteristics were not indicated at all or parts of other key constructs were missing. This led to a final sample size of N = 365. This included 21 cases for which the actual chronological age is unknown.

7.1.4.2 Test for and Treatment of Outliers

Extreme outliers can significantly impact the reliability of statistical analysis.[480] Illegitimate outliers in the data should be eliminated, but identification of outliers and assessment of whether or not they are legitimate is difficult, if not impossible.[481] Methods to keep legitimate outliers, like transformation and truncation, are highly debated among researchers. In this research, these methods will not be used and outliers will be removed from the sample. In order to identify outliers, the non-recursive procedure described by Selst and Jolicoeur (1994) with the suggested cut-off score of 2.5 SD for sample sizes larger than 100 cases was used. This test led to the identification of 14 cases,[482] which were all removed from the final sample size.

---

[478] 72 respondents only clicked the link, 27 respondents read the introduction but did not provide any answers, 49 respondents provide only answers to the first few questions and answered < 20 % of the survey.

[479] 22 of the remaining 405 cases.

[480] Cf. Agresti & Finlay 1997.

[481] Cf. Osborne & Overbay 2008, pp. 206f.

[482] Most outliers were removed due to extreme use experience, i.e., on average more than 120 days per year. Use experience above that threshold indicates that camping and caravanning might not be evaluated as a recreational activity but is an everyday activity in real life. Therefore, the

Therefore, the final sample size used for all following statistical analyses comprised 351 cases, including 18 cases for which the chronological age is unknown. This decreases the sample size for analyses where chronological age matters to N = 333.

For the remaining missing data in the sample, pairwise deletion was conducted and no imputation algorithms were used.

7.2 Findings Regarding Silver Market User Innovators

7.2.1 Results of Descriptive Analysis of Survey Results

The following subchapter aims to provide an overview of the characteristics of the sample and some first insights into factors that lead to innovative behavior via descriptive analysis of the data.

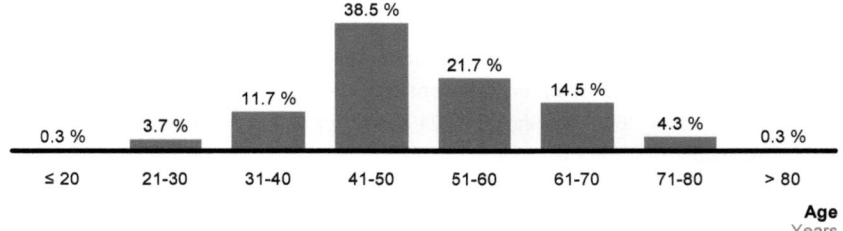

Figure 17: Distribution of Age[483]

The age distribution of the sample is shown in Figure 17 above. The distribution roughly represents a normal distribution, which is slightly right-skewed. There is a sharp increase between the age groups 31 - 40 and 41 - 50. Most responses with 38.5 % came from the age group 41 - 50. This is not surprising because this age group is the largest in the total population. Also, this group has the required financial resources to own a camping vehicle, and the members of this group tend to have children who are old enough to make a trip with basic facilities. Finally, their overall level of fitness is sufficient to independently cater for themselves. The average age of all respondents is 50.0 years and ranges from 19 to 86 years. Participants of the online survey were significantly younger than respondents of the paper-based survey, with an average of 45.9 years compared to 61.0 years. Out of the 333

---

underlying reasons and motivational factors for innovative behavior are probably different. Other outliers were removed if the disposable income was above 81 % or the available time was greater than 16 hours (which is equivalent to the maximal reasonable value).

[483] Own illustration. N_{Age} = 333.

Findings Regarding Silver Market User Innovators

respondents who reported their age, 110 (equal to 33.0 %) were 55 years or older and are attributed to the SiA segment.

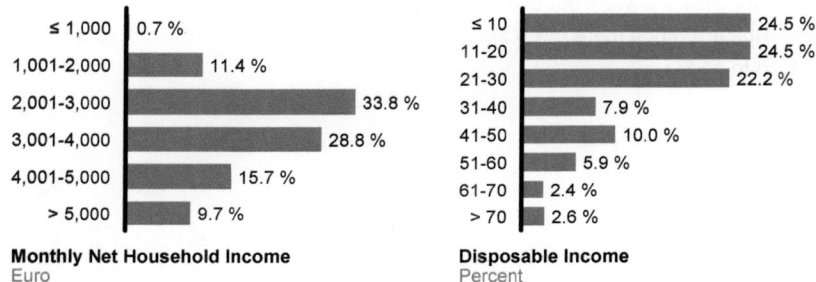

Figure 18: Distribution of Income[484]

The distribution of the monthly net household income peaks in the group of 2,001 - 3,000 EUR, and those with a monthly income between 2,001 EUR and 4,000 EUR represent more than 60 % of all respondents (see Figure 18 above). In contrast, very low incomes (≤ 1,000 EUR) are almost not represented at all. Although camping tourists have a higher average income than the general population (see Table 4 in chapter 5.1.2), the very low share of respondents with low incomes is probably due to the fact that these individuals preferred not to answer this question.[485] Respondents were also asked how large the share of their disposable income is, i.e., which percentage of their monthly net household income is not already spent on compulsory expenses like rent, insurance, food, etc. Figure 18 shows that almost 50 % of respondents can freely dispose of only 20 % or less of their income and only 29 % have more than 30 % of their income at their free discretion. This could lead to differences in innovative behavior, because people with a low disposable income might not have as many resources to innovate. In case they do, they are likely to focus on innovations that help them save money. The data does not support this assumption. The share of innovators varies only slightly among groups, and differences are negligible.[486] An analysis of the type of innovation reveals some

---

[484] Own illustration. $N_{\text{Monthly Net Household Income}}$ = 299; $N_{\text{Disposable Income}}$ = 301.
[485] The N for this question dropped to 299 so that 14.8 % of all respondents did not want to respond to this question. If most of these respondents belong to one of the two lower income groups, the sample distribution would roughly represent the German national average.
[486] Overall innovator share is 42 %. The innovator share of the group with a disposable income "≤ 30 %" is 41 % compared to 44 % for respondents with a higher disposable income. Across all subgroups the innovator share ranges from 38 % to 57 %, except for the group with a disposable income of 51 - 60 %, where the innovator share drops to 17 %. This is considered to be an outlier.

differences between the disposable income groups. There are no differences related to *new functionality*, *compatibility*, and – surprisingly - *cost savings*. Concerning *comfort* innovations, the high disposable income group shows a much higher share (82 % of innovations were labeled comfort innovations, compared to 70 %), while *time savings* are much more often the goal of the low disposable income group (18 % compared to 11 %).

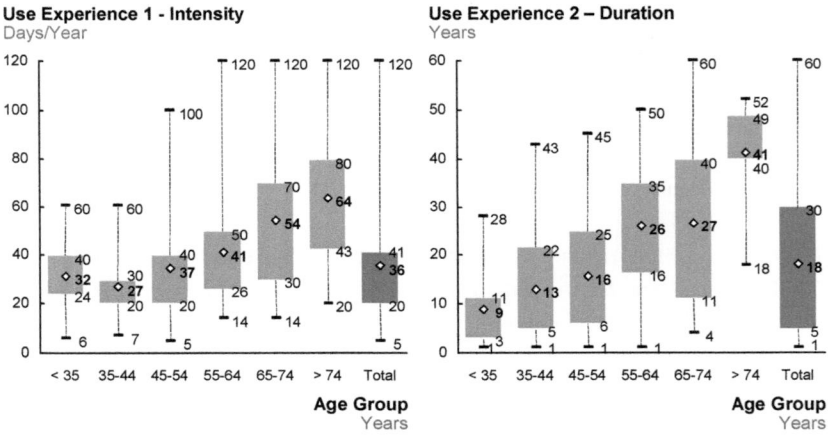

Figure 19: Use Experience Box Plots across Age Groups[487]

The use experience of the respondents was measured with two factors: experience intensity in days per year and experience duration in total years. Both components of use experience increase with age (see Figure 19 above). Respondents spend a considerable amount of their free time camping. The average across all age groups is above 25 days per year, which means that campers spend the majority of their annual vacation time on a campsite. People in the SiMa segment seem to have more time available because the yearly average increases along with the minimum intensity to 14 days / year. This assumption is also supported by the fact that SiAs report to have on average 7.1 hours per day free time, compared to 4.1 hours per day for Non-SiAs. The largest increase can be observed from age groups 55 - 64 to 65 - 74, which can be explained through the additional available time after retirement. Not surprisingly, the total duration of experience increases with chronological age. Nevertheless, novices still exist up to an age of the lower 60s, after which the minimum duration increases strongly. The immediate impact of use experience on

---

[487] Own illustration, N = 333.

the components of lead userness and innovative behavior are detailed in the following chapter.

Differences in disposable time also exist between people with different occupation levels (see Table 10 below). Respondents with a full- or part-time occupation report to have 4.1 to 4.4 hours per day freely available. Respondents who have already retired have more than double that amount of time available.[488] Regardless of less disposable time, the innovator share of full-time employed people is much higher than that of retired people. Also, the comparison of innovator shares across different levels of disposable time shows that innovativeness increases with more time available for the range from 2 h to 8 h per day (see Figure 20 below). Respondents with higher levels are mostly retired, and their innovator share drops considerably. The group with the lowest disposable time (≤ 2 h) has the second highest share of innovators – this is a sign that individuals with limited disposable time are very active and might consider time spent on innovations obligatory, rather than voluntarily.

Occupation	Disposable Time	Innovator Share
Full-Time	4.1 h	49.6 %
Part-Time	4.4 h	36.6 %
Retired	8.9 h	32.4 %
Total	5.1 h	44.7 %

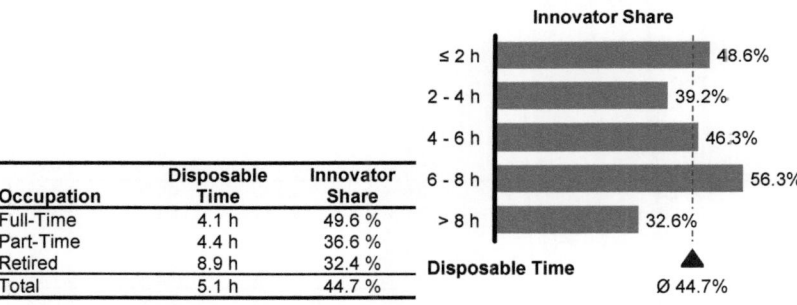

Table 10: Disposable Time and Innovator Share of Occupation Status

Figure 20: Innovator Share of Disposable Time

Generally, one can draw the conclusion that disposable time has a positive impact on innovative behavior, but the impact of retirement (and therefore age) superimposes the effects.

The technical expertise of innovators is considerably higher than non-innovators (see Figure 21 below). Innovators modify their own equipment (TE [1] and TE [4]), they trust that they can help fellow campers (TE [2]), they have solid experience, and they enjoy tinkering (TE [3] and TE [7]). Previous research has shown that technical

---

[488] This result is in line with the latest census on time budget by the German Federal Statistical Office in 2001/2002. Analysis showed that pensioners spend 9 h 43 min on unpaid work, sports, hobbies, and media while full-time employed could only spend 5 h 53 min. The time spent on unpaid work by pensioners (4 h 46 min) was almost twice as high as that by full-time employed (2 h 36 min). Cf. Statistisches Bundesamt Deutschland 2013.

experience is especially important in transforming a simple idea into a working prototype or product.[489]

Table 11: Characteristics of Total Sample, Innovators, and Non-Innovators

		Total	Innovators	Non-Innovators	Δ
Responses		351	157	194	
Age	average	50.0 y	49.9 y	50.2 y	-0.3 y
Cognitive Age	Overall (average)	43.9 y	43.1 y	44.7 y	-1.6 y
	FEEL Age (Ø)	43.3 y	42.0 y	44.2 y	-2.2 y
	LOOK Age (Ø)	46.0 y	45.3 y	46.5 y	-1.2 y
	DO Age (Ø)	42.8 y	41.6 y	43.8 y	-2.1 y
	INTEREST Age (Ø)	43.8 y	43.3 y	44.2 y	-0.9 y
Gender	Male	84.1 %	89.7 %	79.6 %	10.1 %
	Female	15.9 %	10.3 %	20.4 %	-10.1 %
Income	Monthly household inc. (Ø)	3,265 EUR	3,303 EUR	3,238 EUR	65 EUR
	Disposable income (Ø)	28.0 %	28.1 %	27.9 %	0.2 %
Available Time	average	5.1 h	5.0 h	5.3 h	-0.3 h
Education	Secondary school	7.6 %	4.6 %	9.9 %	-5.3 %
	Intermediate school	14.5 %	11.8 %	16.7 %	-4.9 %
	High school	15.1 %	17.1 %	13.5 %	3.6 %
	Apprenticeship	39.0 %	40.1 %	38.0 %	2.1 %
	Master/Diploma	23.0 %	25.0 %	21.4 %	3.6 %
	PhD	0.9 %	1.3 %	0.5 %	0.8 %
Occupation Intensity	Full-time	66.9 %	74.5 %	60.7 %	13.8 %
	Part-time	11.9 %	9.8 %	13.6 %	-3.8 %
	Unemployed	0.6 %	0.7 %	0.5 %	0.2 %
	Retired	20.6 %	15.0 %	25.1 %	-9.9 %
Marital Status	Single	2.6 %	0.6 %	4.1 %	-3.5 %
	In a partnership	13.7 %	16.6 %	11.3 %	3.3 %
	Married	80.1 %	79.0 %	80.9 %	-1.9 %
	Divorced	3.1 %	3.2 %	3.1 %	0.1 %
	Widowed	0.6 %	0.6 %	0.5 %	0.1 %
Use Experience	Intensity (UE [1]) (Ø)	35.9 d/y	36.8 d/y	35.3 d/y	1.5 d/y
	Duration (UE [2]) (Ø)	18.3 y	21.2 y	15.9 y	5.3 y

The sociodemographic characteristics of innovators and non-innovators do not differ much (see Table 11). Both groups have almost identical average ages. Also, income, disposable income, and available time show only very small differences. The majority of the overall sample is male (84 %), because members of the online communities are mostly male and the paper-based survey was often filled out by the head of the household.[490] On top of that, innovators are predominantly male, which is in line with other existing research.[491] Innovators are also more likely to have a better education and are more likely to work full-time and are less likely to be retired. Differences in marital status are negligible. The intensity of use experience does not impact the

---

[489] Cf. Lüthje et al. 2005, pp. 961f.; Lettl et al. 2006, p. 39.
[490] Preparing the camping vehicle and equipment and driving long distances is still a particularly male activity, so the high share of males is not surprising.
[491] Cf. Hippel et al. 2011, p. 28.

likelihood of becoming an innovator, but innovators have, on average, a 33 % longer duration of their experience.

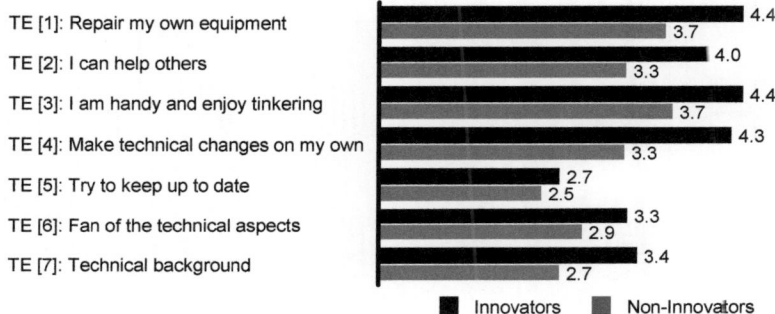

Figure 21: Technical Expertise of Innovators versus Non-Innovators[492]

7.2.2 Findings Regarding Correlations of Chronological Age and Cognitive Age

As stated in chapter 7.1.2.1, cognitive age is believed to reveal more information and to be better suited for conducting social research than chronological age.[493] Several studies have shown that people typically feel eight to thirteen years older than they actually are. The magnitude of the difference differs slightly based on the cultural context, as comparable studies in different countries have shown.[494] To compare the results of existing studies with the characteristics of the study's sample, the analysis of the relationship between chronological age and cognitive age (including its sub-dimensions) is of importance.

---

[492] Own illustration. N = 351.
[493] Cf. Barak & Schiffman 1981, p. 602; Auken & Barry 1995, p. 108.
[494] Results of some selected studies comparing cognitive age and chronological age: 13.5 years younger for Canadians above 55 years (cf. Hubley & Hultsch 1994, p. 425), 10.2 years younger for Australians above 55 years (cf. Hubley & Hultsch 1994, p. 238), 7.9 years younger for Japanese above 50 years (cf. Kohlbacher & Chéron 2011, p. 182). A meta-analysis across cognitive age studies across countries by Barak & Schiffman 1981 showed a minimum difference of 5.4 years and a maximum of 10.9 years for studies with a mean chronological age above 50 years.

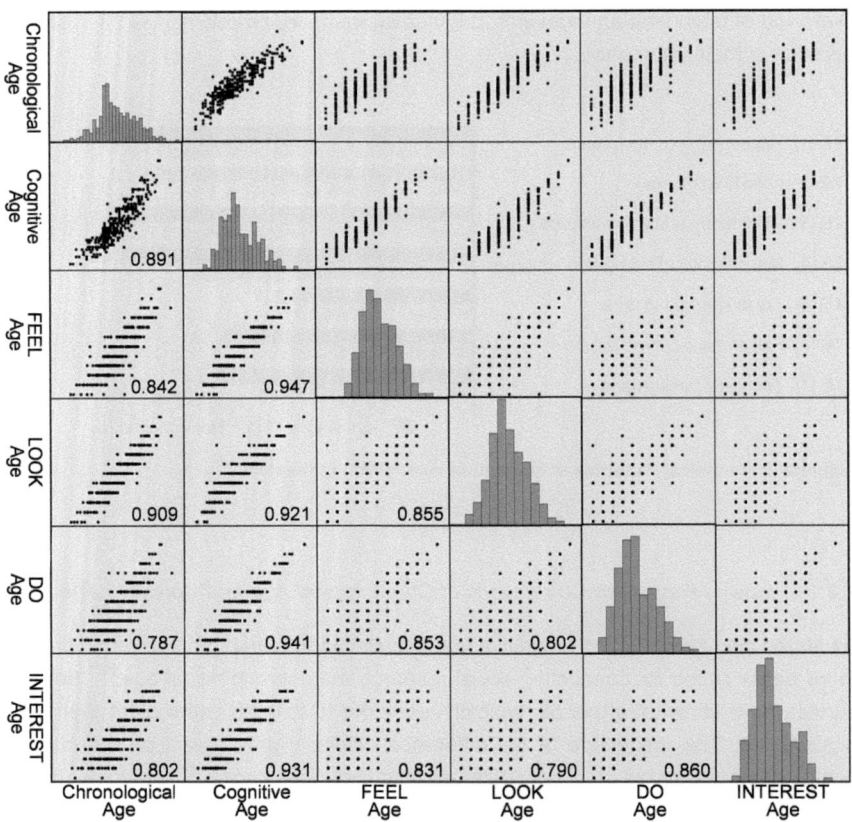

Figure 22: Distribution, Scatter Plots, and Correlations of Age Constructs[495]

Cognitive age was computed for each respondent as the unweighted average of the four sub-dimensions, always using the midpoint of the indicated half-decade.[496] Internal consistency reliability was very high, with a Cronbach's alpha of 0.952. Deletion of any of the four sub-dimensions would not result in a higher Cronbach's alpha. The correlations between the sub-dimension are all high (the lowest correlation coefficient was between LOOK age and INTEREST age at 0.677) and

---

[495] Own illustration. Values in the lower left scatter plots indicate the correlation coefficients between age constructs. All correlation coefficients are significant on a level of p < 0.01.
[496] Cf. Barak & Schiffman 1981, p. 604.

highly significant on a level of p < 0.01. Exploratory factor analysis[497] resulted in only one component with an Eigenvalue > 1, explaining 87.409 % of the variance. The component scores were all well above 0.9, with 0.947 for FEEL age, 0.921 for LOOK age, 0.941 for DO age, and 0.931 for INTEREST age. All other correlation coefficients between the age constructs also verify a strong positive relationship and are highly significant (see Figure 22 above). Chronological age with DO age and LOOK age with INTEREST age show the lowest correlation coefficients, with 0.787 and 0.790 respectively. The highest correlations exist between cognitive age and its sub-dimensions (see above). Chronological age correlates most strongly with LOOK age (0.909) and least with DO age (0.787).

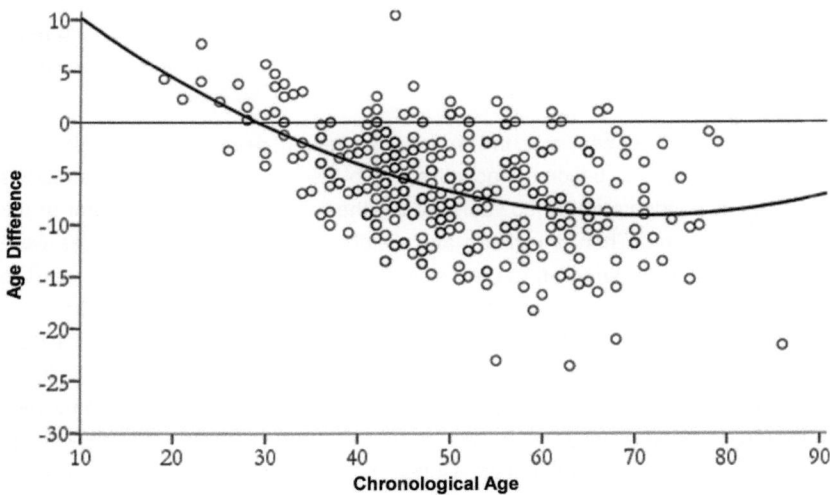

Figure 23: Comparison of Age Difference and Chronological Age[498]

The comparison of characteristics of innovators and non-innovators (see Table 11 above) shows that the two groups differ much more in cognitive age (and its sub-dimensions) than in chronological age. The difference in chronological age is negligible, at 0.3 years. Although the difference in cognitive age is statistically not significant, the difference of the sub-dimensions FEEL age (innovators have a 2.2 years lower FEEL age on average) and DO age (2.1 years lower) is significant on a

---

[497] Exploratory factor analysis on cognitive age was conducted using the principal component analysis without rotation and the Kaiser criterion.
[498] Own illustration. N = 333.
Positive value indicates that the respondent considers him-/herself older than they actually are.

level of $p < 0.10$.[499] FEEL and DO age, therefore, might be suitably explain or even predict innovative behavior.

By subtracting the chronological age from the cognitive age, an age difference can be calculated for each respondent.[500] Negative values for age difference indicate that a person perceives him/herself as being younger than he/she actually is and vice versa. Figure 23 above shows how age difference plots against chronological age in the sample. A quadratic regression fits the data better than a linear regression and explains 23.6 % of the variance.[501] The estimator function is as follows:

Age Difference = 17.021 − 0.735*Chronological Age* + 0.005*Chronological Age²*

Visual inspection of the graphs shows that the estimated age difference is positive below the age of 30 and negative at or above the age of 30, which means that older people perceive themselves younger than they actually are.[502] The age difference increases with age until by the age of 60, when it remains relatively stable at around -8.5 years.[503]

Some researchers suggest that the younger a person perceives him/herself, the more likely he/she is to become innovative.[504] To test this statement for the study's sample, all respondents above 55 years were either assigned to the innovator or the non-innovator group. Then, the mean and distribution of the age differences were compared. As can be seen on the left side of Figure 24 below, innovators tend to perceive themselves as younger than non-innovators. Innovators above 55 years perceive themselves on average 10.0 years younger, while this difference for non-innovators is at just 7.2 years. The mean difference of 2.8 years is highly significant on a level of $p < 0.01$.[505] This difference becomes even more evident if the innovative behavior is examined on a more detailed level. If the development stage of an innovation is taken into account, it shows that the age difference becomes larger the further developed the innovation is. While the age difference is -9.7 years for respondents with only a simple idea, it is -10.2 years for respondents with a working

---

[499] Mean difference of cognitive age = 1.6 (n.s.), mean difference of FEEL age = 2.2 y ($p < 0.10$), mean difference of LOOK age = 1.2 y (n.s.), mean difference of DO age = 2.1 y ($p < 0.10$), mean difference of INTEREST age = 0.9 y (n.s.).

[500] The term *youth age* instead of age difference is also used by some authors. Cf. Barak & Gould 1985, p. 53; Szmigin & Carrigan 2000, p. 517.

[501] Linear regression resulted in an $R^2 = 0.204$. Cubic regression had a slightly higher $R^2 = 0.243$ compared to quadratic regression, but the small gain in explanatory power would come at a much more complicated interpretation of the effect. Therefore, the quadratic regression was selected.

[502] The exact null point of the estimation function is at 28.8 years.

[503] Average age difference is at -6.2 years. The average of the age group 50 - 60 years is at -7.3 years, 60 - 70 years at -8.4 years, and 70 - 80 years at -8.7 years.

[504] Cf. Barak & Schiffman 1981, p. 603; Blau 1973.

[505] Tested through a t-test for equality of means with PASW Statistics 18. N = 110.

prototype and -12.6 years for respondents with an idea that has been commercialized already. Interestingly, respondents, who have stopped at the stage of first drawings or sketches perceive themselves on average only 7.6 years younger, which is not significantly different from the group of non-innovators (-7.2 years).

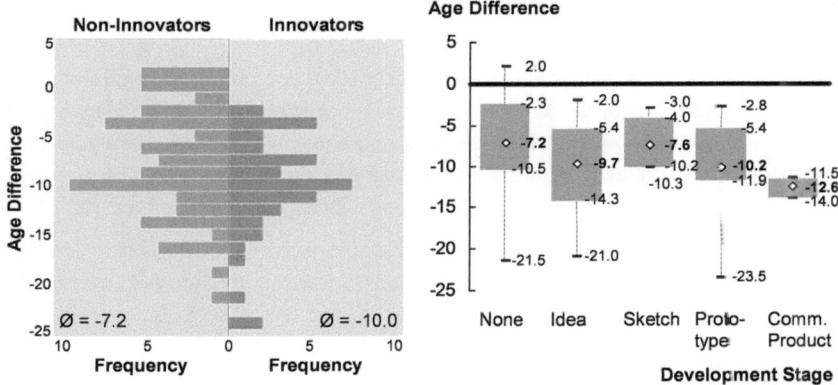

Figure 24: Distribution of Age Differences between Innovators and Non-Innovators above 55 Years[506]

7.2.3 Existence of User Innovators across Age Groups

To evaluate whether user innovators exist across all age groups, the sample was split into 5-year intervals according to chronological and cognitive age. For each 5-year interval, the innovator share was calculated twice: first, based on whether the respondent had stated of having at least an idea for an innovation and, second, based on whether the respondent had at least developed a working prototype. Across all respondents, 45 % had at least an idea and 29 % had developed a working prototype.[507] Compared with previous studies, these numbers are on the upper limit but are still in line with innovator shares for B2C products.[508] As one can see in Figure 25 below, the innovator share based on ideas and chronological age has two peaks. First, there is a general high plateau between the ages of 40 and 64, which is always above 42 % and peaks at 60 % in the age group of 50 – 54. After

---

[506] Own illustration, N = 110.
[507] Differences between age groups regarding the development stage and other innovation characteristics will be detailed in chapter 7.3.
[508] The rate of user innovators was typically 32 - 41 % for ideas and 10 – 26 % for prototypes. See also Table 1 or Franke & Shah 2003; Franke et al. 2006; Lüthje et al. 2002; Lüthje 2004; Tietz et al. 2005.

that, the innovator share drops until a large peak at the oldest age group. A similar pattern exists for the innovator share based on prototypes. The innovator share peaks at the age group of 50 – 54. The high share of innovating ideas in the oldest age group is not repeated for the prototype. The average conversion rate from idea to prototype is 66 %, but it drops to 38 % for people at least 65 years and to 25 % for the oldest age group. Apparently, there exist barriers for people above the age of 65 to transform their ideas into working products.

The distribution of innovator shares based on cognitive age groups is comparable but shows some distinct differences. The innovator share slowly increases until the peak in the age group 40 – 44 and then steadily decreases. People who evaluate their cognitive age to be at least 70 years do not have any innovative ideas. The prototype conversion rate drops to around 20 % for groups with a cognitive age above 60.

Although the visual inspection of the graph would indicate a correlation between cognitive age and innovator share, the correlation analysis did not yield statistically significant results.[509]

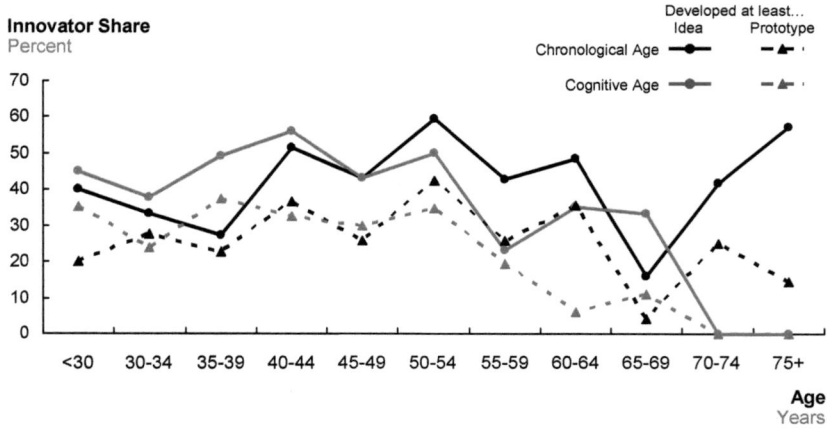

Figure 25: Innovator Shares across Age Groups[510]

The innovator share based on having at least an idea in the SiMa segment (all respondents of at least 55 years) was 39 %, compared to 47 % for younger users.

---

[509] Correlation analysis between chronological age respectively cognitive age and innovative behavior based on an idea respectively a prototype was conducted. The non-parametric tests with Spearman's rho and Kendall's tau did not yield significant results, with $p < 0.10$.

[510] Own illustration. N = 333 for chronological age and N = 351 for cognitive age.

Findings Regarding Silver Market User Innovators

The respective innovator shares based on prototype development were 23 % and 32 %. The Mann-Whitney U-test was used to test whether these group differences are statistically significant.[511] While the results showed no significant difference for the innovator share based on idea development ($p = 0.193$), the innovator share based on prototype development was significantly lower in the SiMa segment ($p = 0.071$).

The respondents were asked about their last innovation, because it was possible that some years had already passed since then. The true age at innovation might, therefore, be different than their current chronological age. This was determined by subtracting the years that passed since the innovation from the current age of the respondents. Unfortunately, this correction could neither be carried out for non-innovators nor for the cognitive age scores. For non-innovators there simply was no "true age at non-innovation", so the age was not corrected. The cognitive age for innovators could not be corrected because it is unknown (and impossible to measure) how innovators would have perceived themselves five or ten years ago. The group comparison can, therefore, only be conducted for innovators based on their chronological age.

Table 12 below shows the shares of the SiMa segment among user innovators (based on idea and prototype development) with their unadjusted and adjusted age at innovation. While the SiMa segment is slightly underrepresented if the unadjusted age at innovation is taken into account (33 % in total sample compared to 29 % / 26 % among innovators), the difference increases for the adjusted age at innovation (18 % / 14 %). Part of this effect is due to the impossible adjustment of age in the group of non-innovators. Although the true effect might be much smaller, it is obvious that the true age at innovation should be taken into account whenever possible.

Table 12: Comparison of Silver Market Shares of Innovators Considering the Adjusted Age at Innovation

Minimum Development Stage	Age at Innovation	< 55 years		≥ 55 years		Total Count
		Count	Share	Count	Share	
Idea	Unadjusted	104	71 %	43	29 %	147
	Adjusted	120	82 %	27	18 %	
Prototype	Unadjusted	72	74 %	25	26 %	97
	Adjusted	83	86 %	14	14 %	
Total sample	Unadjusted	223	67 %	110	33 %	333

---

[511] A non-parametric test had to be used because the test for normal distribution of age with the Kolmogorov-Smirnov test was at $p = 0.069$ and therefore slightly above the generally accepted threshold of 0.05.

Some of the innovations regarding the camping equipment are exemplarily described here:

Foldable baby bed, 37 years (age at innovation: 37 years)

Caravans do not have a baby bed which prevents babies from falling out. A foldable frame was constructed which fits on the regular seat bench and acts as a baby crib. If folded away it fits behind the regular cushion.

Source: 540DM, available online at http://wohnwagen-forum. de/index.php?page=Thread& threadID=59753 (02.01.2014)

Transport case for fuel lamp, 45 years (age at innovation: 42 years)

The Coleman fuel lamp is expensive and glass can break if not taken care of. A transport case made out of readily available material (standardized waste pipe) was built, which allows for safe transport and does not require the disassembly of the lamp.

Source: Niels$, available online at www.klappcaravanforum.de/ viewtopic.php?f=154&t=3170 (02/01/2014)

Mini coat rack, 49 years (age at innovation: 49 years)

A mini coat rack, which can be fixed to any stable tent pole and holds up to six jackets, was constructed. It increases storage space and creates storage space where required.

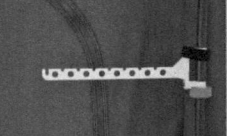

Source: Jasper am Meer, available online at www.klappcaravan forum.de/viewtopic.php?f=165&t=8598 (02/01/2014)

Improved heating, 55 years (age at innovation: 54 years)

To improve the heating for camping trips in colder temperatures a convection heating system was fixed on a board with removable chains. The system is easy to move and install within the caravan and works with any kind of caravan model. Board and chains are foldable and felt seat bases were mounted to the back of the board for scratch protection.

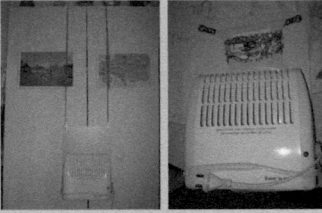

Source: urs_su, available online at www.klappcaravanfor um.de/viewtopic.php?f=60&t=215&start=165 (2/1/2014)

Improved stepladder, 54 years (age at innovation: 42 years)
An ordinary doormat was applied to a metal stepladder to improve comfort and safety.

General improved suitability of daily use, 60 years (age at innovation: 58 years)
Existing equipment and the layout of the caravan was improved upon because manufacturers do not offer specific "senior vehicles" (e.g., vehicles with larger sanitary cabinets) through:
- Installation of aluminum safety edges on all furniture
- Fitting of a full-size fixed bed for more comfort
- Mounting of additional coat hooks
- Construction of a rain shelter for the entrance door to the caravan
- Construction of a transport frame with skateboard wheels for easier transport of the camping toilet

Several adjustments, 61 years (age at innovation: 61 years)
- Installation of a more effective spare tire bracket in a new installation spot to optimize the weight distribution
- Improvement of closing devices of all stowage flaps and cabinet doors
- Installation of baskets to transport dishes more safely and quietly

Satellite dish aid, 76 years (age at innovation: 73 years)
An adjustment aid for mobile satellite dishes was installed with an attached magnetic compass and pre-defined markings for Astra satellite.

7.2.4 Statistical Tests and Bias Treatment

All items were tested for normal distribution using the Kolmogorov-Smirnov-Test and the Shapiro-Wilk-Test.[512] Both tests were significant at a 5 %-level for all indicators, which indicates a deviation from the normal distribution. Therefore, non-parametric tests were used in the following.

As described before, two different modes were used to collect data from respondents: an online survey and a paper-based survey. This setup was used to efficiently invest time and money to collect data while assuring that all age groups would be represented in the final sample. Although mixed mode surveys are the norm in current research, publications on how to reduce measurement errors and secure data quality are scarce.[513] The use of mixed modes during data collection improves coverage and response rates but it can also lead to mode effects on the

---

[512] Cf. Hair et al. 2008, p. 73; Weiber & Mühlhaus 2010, p. 147.
[513] Cf. Leeuw 2005, p. 235.

measurement.[514] To minimize the potential mode effects, the survey was designed identically for both survey types with the same wording for questions, equal answer options, and the same order of questions. Additionally, all guidelines of Dillman, Smyth, and Christian (2009) on the formulation of survey questions for mail and online surveys were applied.[515] In addition to these precautions, the equivalence of data from mixed modes must be tested. The recommended approach to adjust for potential mode effects is to select a random subsample, which is surveyed using both modes.[516] Since this was not feasible for the sample, Leeuw (2005) suggests matching, i.e., "[...] subjects are matched in both modes on important variables, such as age and education, to see if the matched groups are much different."[517] To select appropriate subjects, the 25 % quartile cut-off values and 75 % quartile cut-off values for income, education, and age were computed, and subjects were filtered based on these values. 80 subjects with an income between 2,500 and 3,500 EUR per month, who had at least finished a secondary education and were between 43 and 58 years, were selected. This number was still large enough so that two groups could be compared, and the matching sample was split again by the median age of 50.2 years. Matching sample 1 consisted of 53 subjects, 44 from the online survey and nine from the paper-based survey. Matching sample 2 consisted of 27 subjects, twelve from the online survey and 15 from the paper-based survey. The Mann-Whitney-U Test and the Kolmogorov-Smirnov test were used to test for differences in the groups (see Appendix 2 for detailed figures). For matching sample 1 (with ages from 43 to 50 years) there were no significant differences between groups, according to the Kolmogorov-Smirnov test on a 5 % significance level, and there were only three minor differences, according to the Mann-Whitney-U test. Differences were indicated for TE [1], IB [1], IB [2], and FEEL age. For the slightly older matching sample 2 (with an age from 51 to 58 years), there were a few more differences between groups on a 5 % significance level: TE [1], TE [3], TE [4], IB [2], FEEL age, and job.

Except for the differences in FEEL age, the differences do not indicate a direct mode effect on measurement, and it can be assumed that the measurements of the online survey and the paper-based survey are equivalent. A higher self-evaluation of technical expertise by subjects from the online survey is not surprising, since those subjects have mastered at least one additional technology: the internet. This effect increases with age, which is in line with research on technology acceptance and

---

[514] Cf. Leeuw 2005, p. 238.
[515] Cf. Dillman et al. 2009, pp. 151ff.. A brief overview on guidelines can be found on pages 230 – 233.
[516] Cf. Leeuw 2005, pp. 248f.
[517] Leeuw 2005, p. 249.

age.[518] The impact of online affinity on innovative behavior will be discussed in chapter 8.

Self-reported data is one of the most widely used data collection processes in the social sciences.[519] Although it is widely applied, some scholars criticize its unconstrained application because the use of a single data collection method can lead to a common method bias. This common method bias can be present if data for two variables is gathered from the same source due to shared covariance.[520] To control for common method bias, one can apply procedural and statistical remedies.[521]

Procedural remedies relate to the study design. Since measures of the dependent and independent variables could not be obtained from different sources and a temporal or geographical separation of measurements was not feasible, the other recommendations by Podsakoff et al. (2003) were followed. Most questions included in the survey were formulated as concretely as possible and did not focus on abstract or vague concepts like innate attitudes. The language of questions was kept simple, specific, and concise. Additionally, different anchoring points of scales for dependent and independent variables were used whenever possible, and respondents were granted full anonymity.[522]

Two statistical remedies were applied to assess whether common method bias was an issue in the sample data. First, Harman's single-factor test was applied. All variables of the structural model were loaded into an exploratory factor analysis to check the fit of a single factor solution. The variance extracted for the single factor solution using principal-component-analysis was well below 50 % (29.31%). Five factors with Eigenvalues > 1 were extracted. Therefore, it can be concluded that there does not exist one single factor that accounts for the majority of the covariance among variables.[523] Second, the effects of a single unmeasured latent method factor were controlled for. Using this method, a new first-order factor with all measures as indicators is added to the structural model. All items now load on their theoretical construct and the latent common method factor. Through this method, the variance is partitioned into trait variance, method variance, and random error.[524] Using PLS, it is not possible that one indicator loads on more than on variable, so the procedure

---

[518] Cf. Fisk et al. 2009, pp. 5f.
[519] Cf. Bortz & Döring 2009, pp. 252ff.
[520] Cf. Podsakoff & Organ 1986, p. 533.
[521] Cf. Podsakoff et al. 2003, pp. 887ff.
[522] Cf. Podsakoff et al. 2003, pp. 887f.
[523] Cf. Podsakoff et al. 2003, p. 889.
[524] Cf. Podsakoff et al. 2003, p. 891.

introduced by Liang et al. (2007) should be followed. They proposed to convert each indicator into a single-indicator construct. By doing so, all previous first-order constructs become second-order constructs.[525] In order to interpret the test results, use experience and innovative behavior must be measured reflectively instead of formatively. The theoretical model was compared with both constructs being specified reflectively and formatively. This comparison showed that the specification does not influence test results and that it can be measured reflectively for this test.[526]

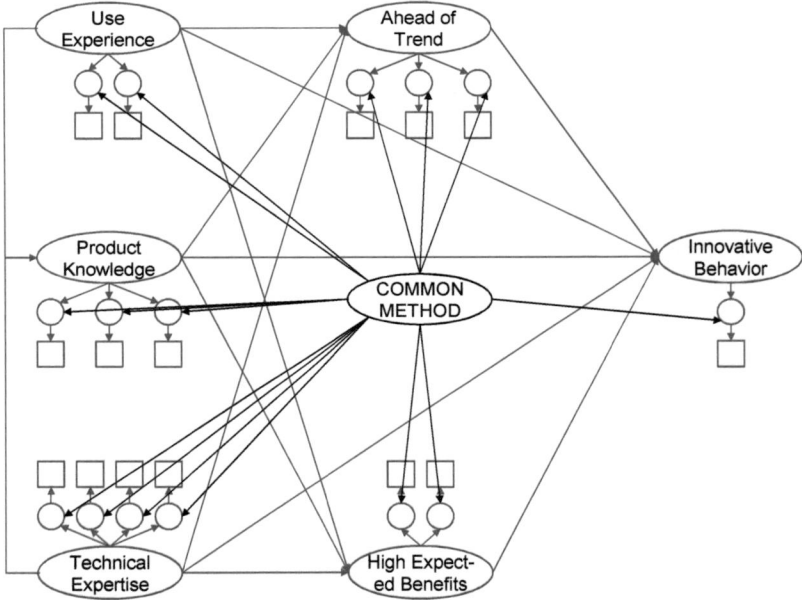

Figure 26: PLS Model for Evaluation of Common Method Bias[527]

According to Williams, Edwards, and Vandenberg (2003), common method bias can be evaluated by examining the statistical significance of the method factor loadings. This is accomplished by comparing the squared substantive and method factor loadings, and by comparing the substantive outer loadings of the model with and without the method factor.[528] Table 13 below shows the results of the model

---

[525] Cf. Liang et al. 2007, pp. 85ff.
[526] This procedure has been applied in previous studies. Cf. Schweisfurth 2013, pp. 95f.
[527] Own illustration.
[528] Cf. Williams et al. 2003, p. 916.

estimations for the common method bias. Evaluation of the method factor loadings shows that only four of the method factor loadings are significant at the level of $\alpha \leq 0.01$, whereas all 14 substantive factor loadings are significant.[529] The squared factor loadings indicate the variance which is based either on the construct or on the method factor. Comparison of the squares yields an average of the explained variance of 0.739 for the constructs and only 0.014 for the method factor.[530] Comparison of the values of the substantive outer loadings in the models with and without the method factor shows that introduction of the method factor does not significantly impact the substantive outer loadings. On average, the loadings stay constant at 0.852.

The evaluation of these two statistical tests suggests that common method bias is not a concern and that the sample can be further analyzed.

Table 13: Evaluation of Common Method Bias

Construct	Indicator	FULL MODEL		FULL MODEL WITH COMMON METHOD FACTOR					
		Substantive Outer Loading λc	T-Value	Substantive Outer Loading λc	T-Value	λc^2	Method Factor Loading λm	T-Value	λm^2
Ahead of Trend	LU [1]	0.854	46.321	0.899	35.309	0.809	-0.080	2.477	0.006
	LU [2]	0.858	48.356	0.890	39.265	0.791	-0.056	1.700	0.003
	LU [4]	0.722	22.253	0.635	14.673	0.404	0.154	3.634	0.024
High Exp. Benefits	LU [5]	0.894	75.797	0.863	52.189	0.744	0.069	2.459	0.005
	LU [6]	0.881	58.992	0.914	46.214	0.834	-0.073	2.377	0.005
Use Experience	UE [1]	0.773	26.363	0.785	30.012	0.616	-0.138	3.198	0.019
	UE [2]	0.804	50.453	0.793	43.317	0.629	0.129	3.512	0.017
Product Knowledge	PK [1]	0.760	25.303	0.877	21.921	0.769	-0.180	3.731	0.032
	PK [2]	0.828	37.497	0.853	23.546	0.727	-0.038	0.889	0.001
	PK [3]	0.787	32.664	0.654	15.329	0.428	0.203	4.241	0.041
Technical Expertise	TE [1]	0.910	63.639	0.958	22.187	0.917	-0.054	1.119	0.003
	TE [2]	0.873	49.979	0.713	12.032	0.508	0.182	3.095	0.033
	TE [3]	0.901	66.332	1.025	21.714	1.051	-0.141	2.599	0.020
	TE [4]	0.935	134.444	0.923	25.771	0.852	0.014	0.351	0.000
Inno. Beh.	IB [1]	1.000	0.000	1.000	0.000	1.000	0.000	0.000	0.000
Average		0.852		0.852		0.739	-0.001		0.014

7.2.5 Model Evaluation

In contrast to many other statistical methods, like regression analysis and CB-SEM, PLS does not search for a global optimum. Therefore, there does not exist one global fit criterion to validate or compare the model.[531] Instead, there are several indicators that can be used to evaluate the measurement model and the structural model

---

[529] LU [4], UE [2], PK [1], and PK [3]. T-value ≥ 3.29.
[530] The ratio of 52.7 : 1 shows most of the variance can be attributed to the constructs.
[531] Cf. Henseler et al. 2012b, pp. 267f.

separately. These all need to be incorporated in order to evaluate the quality of the model.[532]

7.2.6 Measurement Model

The measurement model consists of the latent variable (also called a construct) and its indicators. There exist two types of constructs: formative and reflective. A clear distinction between these two types was demanded by Diamantopoulos and Winklhofer (2001), who emphasized that index development with formative indicators is an alternate and sometimes preferable option.[533] Before them, constructs were often measured reflectively without further justification. According to Diamantopoulos and Winklhofer (2001) the key difference between these measurement approaches is that "*[...] whereas reflective indicators are essentially interchangeable (and therefore the removal of an item does not change the essential nature of the underlying construct), with formative indicators, 'omitting an indicator is omitting a part of the construct'.*"[534] Criteria for the distinction between the measurement approaches are the causal priority between construct and indicators[535], whether indicators represent consequences or causes of the construct[536], or whether all indicators and the construct change consistently in a similar manner[537]. Due to the different characteristics of the reflective and formative measurement approaches, both have to be evaluated differently. This is done in the following chapters.

7.2.6.1 Evaluation of Reflective Constructs

There are several generally accepted criteria for the evaluation of reliability and validity of reflective constructs in the literature. They are typically divided into criteria of the first and second generation.[538] Criteria of the first generation include reliability and validity tests based on overall and construct-specific exploratory factor analysis. Criteria of the second generation test for internal consistency reliability, indicator reliability, convergence reliability, and discriminant validity through the application of confirmatory factory analysis and SEM.[539]

---

[532] Cf. Henseler et al. 2012b, pp. 269ff.; Hair et al. 2011, pp. 144ff.
[533] Cf. Diamantopoulos & Winklhofer 2001.
[534] Diamantopoulos & Winklhofer 2001, p. 271.
[535] Cf. Diamantopoulos & Winklhofer 2001, p. 274.
[536] Cf. Rossiter 2002.
[537] Cf. Chin 1998a, p. 9.
[538] Cf. Weiber & Mühlhaus 2010, pp. 103ff.; Fitzen 2011, pp. 152f.
[539] Cf. Weiber & Mühlhaus 2010, p. 104; Henseler et al. 2009, pp. 319f.; Henseler et al. 2012b, p. 269.

7.2.6.1.1 Exploratory Factor Analysis

The exploratory factor analysis (EFA) aims to "[...] define the underlying structure among the variables in the analysis."[540] Through the application of the EFA, the researcher can test the one-dimensionality of a factor and confirm the theoretically postulated relationship between indicators and their underlying factors.[541] Therefore, separate EFAs for each factor were conducted, followed by an EFA including all confirmed indicators simultaneously. Of specific interest was whether the EFA would indicate that *lead userness* was well represented by one common factor or whether it should be split up into its two sub-dimensions.

As most researchers suggest, the principal component analysis was used as the extraction method and a promax rotation.[542] To determine the appropriateness of the EFA, the *Kaiser-Meyer-Olkin* (KMO) measure of sampling adequacy and *Bartlett's test of sphericity* were applied. The KMO measure indicates how much the indicators correlate with each other and should not be smaller than 0.6.[543] Bartlett's test of sphericity tests whether sufficient correlations exist between the indicators. The null hypothesis should be rejected at least on a significance level < 0.05.[544] On the indicator level, the *measure of sampling adequacy* (MSA) and *communalities* must be analyzed to determine one-dimensionality.[545] The MSA quantifies the amount of intercorrelations of an indicator with the other indicators of a component. The communality indicates the amount of variances that can be explained by the underlying component. Both range from 0 to 1, and indicators with an MSA or communality below 0.5 should be excluded from further analysis.[546] Although with a sample size above 350, a component loading of 0.3 would be statistically sufficient,[547] Hair et al. (2008) recommend only considering loadings of at least 0.5, and this cut-off value was also considered satisfactory for this study. To define the

---

[540] Hair et al. 2008, p. 94.
[541] Cf. Weiber & Mühlhaus 2010, p. 106.
[542] Cf. Hair et al. 2008, pp. 105ff.; Bortz & Schuster 2010, pp. 389ff.
The principal component analysis is most appropriate if the primary objective is data reduction and the maximum portion of explained total variance shall be explained. Hair et al. 2008, pp. 107f.
The promax rotation belongs to the oblique rotation methods. These rotation methods assume that the factors are at least somehow correlated. This is in contrast to orthogonal rotation methods, which assume no correlation between factors at all. An oblique rotation method was chosen because it usually provides more meaningful factors and realistically there are almost no factors which are completely uncorrelated. Hair et al. 2008, pp. 115f.; Weiber & Mühlhaus 2010, pp. 107f.
[543] Cf. Kaiser & Rice 1974, p. 112.
[544] Cf. Dziuban & Shirkey 1974, pp. 358f.; Hair et al. 2008, p. 105.
[545] Cf. Weiber & Mühlhaus 2010, p. 107.
[546] Cf. Weiber & Mühlhaus 2010, p. 107; Hair et al. 2008, p. 105; Backhaus et al. 2011, p. 341.
[547] Cf. Hair et al. 2008, p. 117.

number of extracted components, the Kaiser criterion was applied, which extracts all components with an Eigenvalue greater than 1.[548]

To check the reliability of the constructs Cronbach's alpha, inter-item-correlation (IIC) and the corrected item-to-total-correlation (CITC) were taken into account. Cronbach's alpha "[...] **absolutely** [highlighted in original through italic script] should be the first measure one calculates to assess the quality of the instrument."[549] A widely accepted minimum threshold suggested by Nunnally and Bernstein (1994) requires an alpha of at least 0.7.[550] The IIC calculates the average correlation between indicators and should at least be 0.3.[551] An additional measure is the CITC, which measures the correlation of an indicator with the sum of all other indicators of its theoretical construct.[552] Indicators with a CITC value below 0.5 should be eliminated.[553]

EFA Results for Lead Userness

As mentioned above, one key objective of the EFA was to test whether the lead userness construct could be measured through one factor or whether the two dimensions would call for the distinction of two factors. As Table 14 below shows, the EFA suggested the existence of two distinct components with Eigenvalues > 1. Items LU [1], LU [2], LU [3], and LU [4] formed component 1, while LU [5] and LU [6] formed component 2.

Referring to the content of the items of component 1, one realizes that all items refer to the *ahead of trend* dimension of lead userness while all items of component 2 describe the *high expected benefits* dimension. The two extracted components show a relatively low correlation of 0.288, which also suggests that both components should be measured separately.[554] This distinction also confirms the suggestion by Franke and Shah (2003) to examine the sub dimensions of lead userness separately.[555] Based on these results, the lead userness construct was split into its two sub dimensions for all further analyses.

EFA Results for Ahead of Trend

As can be seen in

---

[548] Cf. Weiber & Mühlhaus 2010, p. 107.
[549] Churchill Jr. 1979, p. 68.
[550] Cf. Nunnally & Bernstein 1994, pp. 264f.
[551] Cf. Robinson et al. 1991, p. 13.
[552] Cf. Weiber & Mühlhaus 2010, p. 112.
[553] Cf. Zaichkowsky 1985, p. 343; Bearden et al. 1989, p. 475.
[554] Cf. Weiber & Mühlhaus 2010, p. 109.
[555] Cf. Franke & Shah 2003, p. 163; Franke et al. 2006, p. 303.

Findings Regarding Silver Market User Innovators 115

Table 15 below, the KMO measure (0.658 > 0.6) and Bartlett's test of sphericity (H_0 is rejected) fulfill the quality criteria, and the data is suitable for conducting a factor analysis. LU [3] shows a communality below 0.3 and a CITC below 0.5. It has the lowest factor loadings of all items, so it should be eliminated. Elimination would also improve Cronbach's alpha, making it 0.743.

EFA Results for High Expected Benefits

Because the construct high expected benefits only contains two items, the KMO measure and the MSA equal exactly 0.500.[556] Bartlett's test of sphericity shows that the data is suitable for conducting a factor analysis. The communalities (0.787) and factor loadings (0.887) are sufficient and Cronbach's alpha is 0.730 and is, therefore, above the required threshold. As such, none of the items needs to be eliminated.

EFA Results for Technical Expertise

As can be seen in

Table 17 below, the KMO measure (0.850 > 0.6) and Bartlett's test of sphericity (H_0 is rejected) fulfill the quality criteria, and the data is suitable for conducting a factor analysis. The analysis shows that TE [5] and TE [6] form a second component. Additionally, the communality of TE [7] at 0.392 is below the threshold of 0.5. Based on these results, all three items were eliminated from the construct and only the first four were kept. This also improved Cronbach's alpha (from 0.857 to 0.925) and inter-item-correlation (from 0.488 to 0.758). Reduction of items to TE [1] - [4] is also in line with Franke, Hippel, and Schreier (2006), the original creator of this measure, who eliminated the same items after validity tests.[557]

EFA Results for Product Knowledge

As can be seen in Table 18 below, the KMO measure (0.663 > 0.6) and Bartlett's test of sphericity (H_0 is rejected) fulfill the quality criteria and the data is suitable for a factor analysis. All values for MSA and communality are above the suggested threshold of 0.5, and Cronbach's alpha is slightly above the required value of 0.7. The CITC of PK [1] at 0.493 is slightly below the threshold of 0.5. Nevertheless, PK [1] was kept because the difference is very marginal, and eliminating PK [1] would result in a Cronbach's alpha of 0.644.

---

[556] Since there are only two items, the linear combination of these items is exactly the average.
[557] Cf. Franke et al. 2006, p. 315.

Table 14: Results of Exploratory Factor Analysis for Lead Userness

Kaiser-Meyer-Olkin Measure of Sampling Adequacy		0.658		
Significance of Bartlett's Test of Sphericity		0.000		
Item	MSA	Communality	Component 1	Component 2
LU [1]	0.662	0.672	**0.806**	0.043
LU [2]	0.641	0.723	**0.874**	-0.105
LU [3]	0.829	0.295	**0.564**	-0.097
LU [4]	0.795	0.514	**0.633**	0.199
LU [5]	0.609	0.781	0.070	**0.861**
LU [6]	0.554	0.788	-0.103	**0.912**
Initial Eigenvalue			2.468	1.305
Rotation Sums of Squared Loadings			2.274	1.803
Explained Variance (after Rotation)			62.884 %	

Table 15: Results of Initial Exploratory Factor Analysis for Ahead of Trend

Kaiser-Meyer-Olkin Measure of Sampling Adequacy					0.681		
Significance of Bartlett's Test of Sphericity					0.000		
Item	MSA	Communality	Component 1	CITC	IIC	Cronbach's α	Cronbach's α without item
LU [1]	0.636	0.678	0.823	0.625	0.371	0.711	0.562
LU [2]	0.636	0.699	0.836	0.645			0.545
LU [3]	0.806	0.271	0.521	0.307			0.743
LU [4]	0.799	0.504	0.710	0.473			0.665
Initial Eigenvalue			2.152				
Explained Variance			53.790 %				

Table 16: Results of Exploratory Factor Analysis for High Exptected Benefits

Kaiser-Meyer-Olkin Measure of Sampling Adequacy					0.500		
Significance of Bartlett's Test of Sphericity					0.000		
Item	MSA	Communality	Component 1	CITC	IIC	Cronbach's α	Cronbach's α without item
LU [5]	0.500	0.787	0.887	0.575	0.575	0.728	-
LU [6]	0.500	0.787	0.887	0.575			-
Initial Eigenvalue			1.575				
Explained Variance			78.728 %				

Table 17: Results of Initial Exploratory Factor Analysis for Technical Expertise

Kaiser-Meyer-Olkin Measure of Sampling Adequacy		0.850		
Significance of Bartlett's Test of Sphericity		0.000		
Item	MSA	Communality	Component 1	Component 2
TE [1]	0.893	0.807	**0.917**	-0.047
TE [2]	0.903	0.733	**0.804**	0.111
TE [3]	0.865	0.800	**0.917**	-0.058
TE [4]	0.840	0.870	**0.931**	0.005
TE [5]	0.682	0.848	-0.116	**0.963**
TE [6]	0.773	0.789	0.152	**0.814**
TE [7]	0.914	0.392	**0.621**	0.011
Initial Eigenvalue			4.054	1.184
Rotation Sums of Squared Loadings			3.910	2.253
Explained Variance (after Rotation)			74.830 %	

Table 18: Results of Exploratory Factor Analysis for Product Knowledge

Kaiser-Meyer-Olkin Measure of Sampling Adequacy			0.663				
Significance of Bartlett's Test of Sphericity			0.000				
Item	MSA	Commu-nality	Compo-nent 1	CITC	IIC	Cronbach's α	Cronbach's α without item
PK [1]	0.686	0.595	0.771	0.493	0.440	0.702	0.644
PK [2]	0.633	0.684	0.827	0.567			0.545
PK [3]	0.681	0.602	0.776	0.501			0.633
Initial Eigenvalue			1.882				
Explained Variance			62.725 %				

Table 19: Results of Exploratory Factor Analysis across all Selected Items

Kaiser-Meyer-Olkin Measure of Sampling Adequacy			0.796			
Significance of Bartlett's Test of Sphericity			0.000			
	MSA	Commu-nality	Component 1	Component 2	Component 3	Component 4
Crit. Value	≥ 0.5	≥ 0.5	> 0.5			
Item						
LU [1]	0.658	0.784	-0.010	**0.910**	-0.100	0.012
LU [2]	0.697	0.772	0.003	**0.904**	0.010	-0.133
LU [4]	0.881	0.491	0.044	**0.556**	0.115	0.192
LU [5]	0.654	0.780	-0.012	0.054	0.044	**0.861**
LU [6]	0.579	0.791	-0.009	-0.071	-0.073	**0.916**
TE [1]	0.874	0.837	**0.928**	-0.060	0.004	-0.003
TE [2]	0.898	0.759	**0.820**	0.093	0.065	-0.020
TE [3]	0.852	0.821	**0.933**	-0.041	-0.032	-0.029
TE [4]	0.818	0.870	**0.938**	0.046	-0.075	0.028
PK [1]	0.771	0.687	-0.080	-0.114	**0.887**	-0.066
PK [2]	0.744	0.701	-0.091	0.173	**0.794**	0.023
PK [3]	0.833	0.608	0.284	-0.068	**0.644**	0.034
Initial Eigenvalue			4.385	1.884	1.462	1.170
Rotation Sums of Squared Loadings			3.862	2.568	2.560	2.034
Explained Variance			74.177 %			

EFA Results across all Reflective Constructs

Table 19 and Table 20 show the combined first generation quality criteria for all selected final items. As can be seen in

Table 19 above, the KMO measure (0.796 > 0.6) and Bartlett's test of sphericity (H_0 is rejected) fulfill the quality criteria, and the data is suitable for conducting a factor analysis. The lowest MSA score is at 0.579 and is, therefore, well above the suggested threshold of 0.5. The same applies for the communalities, except for LU [4]. Its communality, at 0.491, is slightly below the suggested threshold of 0.5, as is the corrected item-to-total-correlation, at 0.448. LU [4] was still kept because overall quality criteria for the construct *ahead of trend* (see Table 20 below) were within the suggested ranges and measuring a construct with three items is preferred, rather than with just two. The EFA across all selected items extracted four components with Eigenvalue > 1 and a total explained variance of 74.177 %. All items were assigned to their respective, theoretically derived constructs.

Cronbach's alphas of the reflective constructs range from 0.702 to 0.925 and are therefore all in an acceptable range. Inter-item-correlations range from 0.440 to 0.758 and are all well above the required critical value of 0.3 (see Table 20 below).

Table 20: Final Results of EFA for all Reflective Constructs (measured separately)

Construct	Item	MSA	Communality	Comp. Loading	CITC	IIC	Cronbach's α	Cronbach's α without item
Critical Value		≥ 0.5	≥ 0.5	> 0.5	≥ 0.5	≥ 0.3	≥ 0.7	≤ Cronbach's α with item
Ahead of Trend	LU [1]	0.609	0.738	0.859	0.648	0.489	0.744	0.563
	LU [2]	0.606	0.745	0.863	0.655			0.552
	LU [4]	0.798	0.504	0.710	0.448			0.790
High Exp. Benefits	LU [5]	0.500	0.787	0.887	0.575	0.575	0.728	-
	LU [6]	0.500	0.787	0.887	0.575			-
Technical Expertise	TE [1]	0.867	0.829	0.910	0.837	0.758	0.925	0.901
	TE [2]	0.873	0.759	0.871	0.776			0.919
	TE [3]	0.830	0.814	0.902	0.824			0.904
	TE [4]	0.801	0.874	0.935	0.878			0.885
Product Knowledge	PK [1]	0.686	0.595	0.771	0.493	0.440	0.702	0.644
	PK [2]	0.633	0.684	0.827	0.567			0.545
	PK [3]	0.681	0.602	0.776	0.501			0.633

7.2.6.1.2 Indicator Reliability

Indicator reliability measures how much of the variance of an item is explained by the variance of the causing latent variable.[558] It is usually assessed by the magnitude of the outer loading and its significance. The outer loading should exceed the value of 0.7, and the relationship should be highly significant.[559] Items with a loading below 0.4 should be deleted. For items with loadings between 0.4 and 0.7, the effect of the deletion of the item on composite reliability (see chapter 7.2.6.1.3 below) and average variance extracted (see chapter 7.2.6.1.4 below) should be analyzed. If the effect is positive, the item should be deleted.[560]

---

[558] Cf. MacKenzie et al. 2011, p. 314.
[559] Cf. Hair et al. 2011, p. 145.
[560] Cf. Hair et al. 2013, p. 104.

The analysis of the outer loadings of all items confirms the results of the exploratory factor analysis (see Table 21 below).[561] The outer loadings of all selected items for the structural model are sufficient and are highly significant (see Table 22 below).

7.2.6.1.3 Internal Consistency Reliability

The internal consistency reliability specifies whether all indicators of a construct measure the same thing and the degree to which they are interrelated.[562] It is traditionally measured via Cronbach's alpha, which measures the average correlation between all items, and for which a minimum value of 0.7 is required.[563] More suitable in the context of PLS-SEM is the application of composite reliability, because (in contrast to Cronbach's alpha) composite reliability does not assume equal reliability of all indicators.[564] Composite reliability was measured through Dillon-Goldstein's rho and a critical value of at least 0.7 for satisfactory results was applied.[565] As can be seen in Table 22 below, all internal consistency reliability measures for reflective constructs with the selected items are within satisfactory ranges.

7.2.6.1.4 Convergent Validity

The convergent validity describes how much variance the items of a construct share with the overall construct and, therefore, the extent to which these items converge.[566] Convergent validity can be confirmed by using independent measurement procedures.[567] Due to a limited access to respondents and restrictions on survey length, a multi-method approach was not feasible. As an alternative, the average variance extracted (AVE) is typically applied in the realm of PLS-SEM to assess convergent validity. Typically a minimum value of 0.5 is required, so that the construct captures at least 50 % of the variance and less than 50 % of the variance is due to measurement or random error.[568] For this study, the minimum value was defined at 0.6. As can be seen in Table 22 below, AVE is between 0.624 and 0.818. Convergent validity is therefore established for all reflective constructs.

---

[561] The outer loading of TE [6] is 0.706 in the initial model with all items. It drops to 0.650 if TE [5] and TE [7] are removed from the model. Because of this drop below the critical value and the results of the previous exploratory factor analysis, TE [6] is also removed from the final measurement model and the construct technical expertise is measured through the items TE [1] - [4].
[562] Cf. Hair et al. 2008, p. 634.
[563] Cf. MacKenzie et al. 2011, p. 314; Hair et al. 2012, p. 424. See also chapter 7.2.6.1.1.
[564] Cf. Hair et al. 2011, p. 145; Hair et al. 2012, p. 424.
[565] Cf. Esposito Vinzi et al. 2010, pp. 50f.; Hair et al. 2011, pp. 145f.
[566] Cf. Hair et al. 2008, p. 689.
[567] Cf. Campbell & Fiske 1959, p. 81.
[568] Cf. Fornell & Larcker 1981, p. 46; Hair et al. 2011, p. 145.

7.2.6.1.5 Discriminant Validity

While convergent validity describes only whether the items of a construct converge, discriminant validity assesses whether the relationship of items to their theoretically derived construct is higher than to any other construct in the model.[569] There exist two measures that are typically applied in PLS research to test for discriminant validity: examination of cross loadings and the Fornell-Larcker criterion.[570] The Fornell-Larcker-criterion measures discriminant validity on a construct level and requires the square of the construct correlations to be smaller than the average variance extracted.[571]

Table 24 compares the positive square root of the average variance extracted with the construct correlations, which provides the exact same results. The highest construct correlation is 0.414, while the lowest square root of average variance extracted is 0.790, so the Fornell-Larcker criterion is fulfilled. The examination of cross loadings tests for discriminant validity occurs on the item level. According to Chin (1998b), the loading of an item on its theoretically derived construct should be higher than the loadings on any other latent variable.[572] Table 23 shows that this requirement is fulfilled and that discriminant validity is also confirmed on the item level.

---

[569] Cf. Campbell & Fiske 1959, p. 84; Hair et al. 2008, p. 689; MacKenzie et al. 2011, pp. 323f.
[570] Cf. Hair et al. 2011, p. 145; Hair et al. 2012, p. 423; Henseler et al. 2009, pp. 299f.
[571] Cf. Fornell & Larcker 1981, p. 46.
[572] Cf. Chin 1998b, p. 321.

Table 21: Evaluation of All Items of Reflective Constructs of Measurement Model

		INDICATOR RELIABILITY		INTERNAL CONSISTENCY RELIABILITY		CONVERGENT VALIDITY
		Standardized Outer Loading λ	T-Value	Dillon-Goldstein's ρ	Cronbach's α	Average Variance Extracted
Critical Value		λ ≥ 0.7	≥ 1.96 : p < 0.05 ≥ 2.58 : p < 0.01 ≥ 3.29 : p < 0.001	ρ ≥ 0.7	α ≥ 0.7	AVE ≥ 0.6
Construct	**Item**					
Ahead of Trend	LU [1]	0.799	25.564	0.813	0.702	0.532
	LU [2]	0.807	25.828			
	LU [3]†	0.461†	5.628			
	LU [4]	0.790	32.400			
High Exp. Benefits	LU [5]	0.927	72.415	0.878	0.732	0.784
	LU [6]	0.842	27.619			
Technical Expertise	TE [1]	0.835	33.420	0.902	0.870	0.574†
	TE [2]	0.845	50.324			
	TE [3]	0.823	33.659			
	TE [4]	0.884	58.825			
	TE [5]†	0.560†	11.951			
	TE [6]††	0.706	21.865			
	TE [7]†	0.583†	14.627			
Product Knowledge	PK [1]	0.700	14.939	0.830	0.702	0.621
	PK [2]	0.828	33.766			
	PK [3]	0.830	32.942			

Cases: 351; Samples: 5,000

† Items considered for deletion because of insufficient critical values.
†† Item considered for deletion because after deletion of TE [5] and TE [7], outer loading dropped to 0.650. Additionally, EFA showed that it loads on a different component than TE [1] - [4] (see chapter 7.2.6 1.1 above).

Table 22: Evaluation of Selected Items of Reflective Constructs of Measurement Model

		INDICATOR RELIABILITY		INTERNAL CONSISTENCY RELIABILITY		CONVERGENT VALIDITY
		Standardized Outer Loading λ	T-Value	Dillon-Goldstein's ρ	Cronbach's α	Average Variance Extracted
Critical Value		λ ≥ 0.7	≥ 1.96 : p < 0.05 ≥ 2.58 : p < 0.01 ≥ 3.29 : p < 0.001	ρ ≥ 0.7	α ≥ 0.7	AVE ≥ 0.6
Construct	Item					
Ahead of Trend	LU [1]	0.810	26.250	0.851	0.741	0.655
	LU [2]	0.817	25.309			
	LU [4]	0.799	29.470			
High Exp. Benefits	LU [5]	0.939	51.978	0.875	0.730	0.778
	LU [6]	0.821	20.836			
Technical Expertise	TE [1]	0.906	58.440	0.947	0.926	0.818
	TE [2]	0.888	63.885			
	TE [3]	0.891	55.876			
	TE [4]	0.932	121.972			
Product Knowledge	PK [1]	0.701	15.030	0.832	0.702	0.624
	PK [2]	0.821	29.327			
	PK [3]	0.836	29.557			

Cases: 351; Samples: 5,000

Table 23: Cross Loadings

	AoT	HEB	TE	PK	IB[a]	UE[a]
LU [1]	**0.797**	0.258	0.210	0.210	0.045	-0.012
LU [2]	**0.806**	0.168	0.215	0.285	0.064	0.046
LU [4]	**0.817**	0.313	0.271	0.307	0.243	0.123
LU [5]	0.338	**0.928**	0.226	0.185	0.419	0.018
LU [6]	0.200	**0.838**	0.162	0.085	0.321	0.048
TE [1]	0.213	0.192	**0.908**	0.387	0.359	0.105
TE [2]	0.325	0.199	**0.885**	0.432	0.374	0.114
TE [3]	0.218	0.168	**0.891**	0.354	0.299	0.050
TE [4]	0.290	0.246	**0.934**	0.347	0.424	0.072
PK [1]	0.161	0.041	0.214	**0.704**	0.183	0.182
PK [2]	0.364	0.165	0.265	**0.809**	0.172	0.192
PK [3]	0.261	0.154	0.463	**0.844**	0.274	0.215
IB [1]	0.151	0.412	0.373	0.258	**0.953**	0.187
IB [2]	0.170	0.414	0.408	0.270	**0.982**	0.173
UE [1]	0.001	-0.096	-0.148	0.139	0.002	**0.157**
UE [2]	0.076	0.024	0.081	0.258	0.180	**0.996**

a) Formative construct

Table 24: Discriminant Validity (Fornell-Larcker Criterion)

	AoT	HEB	TE	PK
Ahead of Trend	0.809			
High Expected Benefits	0.316	0.882		
Technical Expertise	0.292	0.225	0.905	
Product Knowledge	0.340	0.165	0.414	0.790

Diagonal: Square root of average variance extracted
Fields below diagonal: Construct correlations

7.2.6.2 Evaluation of Formative Constructs

In formative constructs, the individual items cause the latent construct and *"[...] dropping a measure from a formative-indicator model may omit a unique part of the conceptual domain and change the meaning of the variable, because the construct is a composite of all the indicators."*[573]

There are two constructs in this study that are measured formatively: innovative behavior and use experience. Innovative behavior is measured with a single item measurement.[574] Use experience is measured with two items (UE [1] and UE [2]).

For the evaluation of formative constructs, there do not exist generally accepted criteria in the literature. Some authors argue that an evaluation is not possible at all[575], while others point out that the lack of overall quality standards might lead to a certain ambiguity at the building of constructs.[576] The commonly applied criteria for reflective constructs (e.g., Cronbach's alpha, composite reliability) are inappropriate for formative constructs. Additionally, reliability, convergent validity, and discriminant validity cannot be used to assess the construct's quality.[577] Instead, one needs to (1) discuss the face and content validity, (2) assess each indicator's weight and loading, and (3) check for potential multicollinearity issues.

1) Face and Content Validity

Face validity and content validity are mostly used interchangeably.[578] They describe the extent to which the content and wording of an item is consistent with the definition of the construct. It cannot be tested statistically and is based on the researcher's judgment.[579] The researcher should evaluate the concrete formulation of the items as well as their applicability in practical situations, i.e., Does it correspond to the respondents reading level? Is it clear and easy of use?[580]

Innovative behavior uses a single item measurement. The main model refers to whether a respondent has innovated on his camping equipment before (see chapter 7.1.2.1). This item has been successfully applied in previous

---

[573] MacKenzie et al. 2005, p. 712.
[574] In the main structural model, it is measured as innovative action (IB [1]). Further analyses include the development stage (IB [2]) as an alternative – but still single item – measurement. See also chapter 7.1.2.1.
[575] Cf. Albers & Hildebrandt 2006, p. 13.
[576] Cf. Homburg & Klarmann 2006, p. 731.
[577] Cf. Hair et al. 2012, p. 423; Diamantopoulos 2006, p. 11; Hair et al. 2011, p. 146.
[578] Cf. Netemeyer et al. 2003, p. 12.
[579] Cf. Hair et al. 2008, p. 689.
[580] Cf. Netemeyer et al. 2003, pp. 72ff.

research,[581] and the formulation of the item has been checked by experts during the pre-test. Use experience is measured using two indicators that take into account the frequency and length of use. The construct *use experience* has been used in several studies before and has been shown to be a useful measurement.[582]

Therefore, it was assumed that face and content validity is established.

2) Indicator Weights and Loadings

The analysis of the outer weights and outer loadings of an item is required to evaluate the absolute and relative importance. Absolute importance is assessed through the outer loadings and relative through the outer weights. Additionally, bootstrapping is carried out to assess the significance of the items.[583]

Since innovative behavior is only measured with a single item, outer loading and outer weight are equal to 1. The values for use experience are shown in Table 25 below. As can be seen, the outer weight for frequency of use (UE [1]) is not significant, and the respective loadings and weights are much lower than the total length of use (UE [2]). The outer loading of UE [1] is slightly below the critical value of 0.5, but it is significant on a level of $p < 0.1$. UE [2] shows high absolute and relative importance, which is also highly significant. Although Hair, Ringle, and Sarstedt (2011) suggest questioning the theoretical relevance if weights are not significant, frequency of use was still kept in the data. The intensity of an activity also defines the depth of gained experiences.[584] In the case of camping, the intensity can be defined by extreme ways of camping (e.g., winter camping) and by the amount of time a person spends on camping per year. Since extreme camping is practiced by few individuals and almost does not apply to the SiMa segment, the intensity of gained experiences is approximated via the frequency of use.[585] Additionally, the PLS algorithm is capable to account for the imbalance of importance of the two items and can assign appropriate weights during the calculation.

3) Multicollinearity

Indicators of formative constructs should only marginally overlap, so formative constructs need to be checked for multicollinearity. In PLS-SEM, the variance inflation factor (VIF) is usually used to assess multicollinearity. The VIF for both items

---

[581] Cf. Franke & Shah 2003; Lüthje 2000; Lüthje et al. 2005; Franke et al. 2006.
[582] Cf. Schweisfurth 2013; Lüthje et al. 2005; Lüthje 2004.
[583] Cf. Hair et al. 2011, p. 145.
[584] Tietz et al. 2005, p. 331.
[585] Frequency as a proxy of the intensity of usage has been applied in many lead user studies. See for example Lüthje et al. 2005; Lüthje 2004; Schweisfurth & Raasch 2012; Schreier & Prügl 2008.

of use experience is 1.048, which is well below the suggested cut-off value of 5.[586] Since use experience consists of only two items, correlation analysis was also applied. The Pearson correlation coefficient is low, at 0.245.[587] Therefore, multicollinearity does not prevail.

Table 25:   Evaluation of Formative Measure "Use Experience"

	Outer Loading	Outer Loading's T-Value	Outer Weight	Outer Weight's T-Value	Variance Inflation Factor	Correlation
Critical Value		≥ 1.96 : p < 0.05 ≥ 2.58 : p < 0.01 ≥ 3.29 : p < 0.001		≥ 1.96 : p < 0.05 ≥ 2.58 : p < 0.01 ≥ 3.29 : p < 0.001	VIF < 5	
Item						
UE [1]	0.421	1.855	0.192	0.750	1.048	0.245***
UE [2]	0.983	9.703	0.936	6.570	1.048	

Bootstrapping with 351 cases and 5,000 samples
*** Pearson correlation significant with p < 0.01 (2-tailed)

7.2.7 Evaluation of Structural Model – Determinants of Innovative Behavior

Since the PLS algorithm does not solve for a global optimum, there does not exist a global quality criterion to evaluate the fit of the structural model.[588] Instead there are several criteria, which should be analyzed to evaluate the model: level and significance of path coefficients, explained variance, and predictive relevance.[589] The interpretation of the level of path coefficients in the structural model is equivalent to the interpretation of standardized beta coefficients in ordinary least squares regressions.[590] The standardized path coefficients should be at least 0.2 to be considered meaningful.[591] The significance of the path coefficients in PLS is

---

[586] Cf. Hair et al. 2011, p. 145. Henseler et al. 2009, p. 302 suggest that a VIF substantially greater than 1 already indicates multicollinearity, but the presented value of 1.048 is just slightly above that threshold.

[587] Correlation significant on p < 0.01 (2-tailed).

[588] Cf. Henseler et al. 2012b, p. 267 .
There is a global goodness-of-fit (GoF) criterion, introduced by Tenenhaus, Amato, and Esposito Vinzi (2004), which is calculated as the geometric mean of the average variance explained of the measurement model (via communalities) and the average variance explained of the structural model (via R^2). A critique of the GoF is that it only works properly with reflective measurement models, is not defined for single item constructs, and ignores the complexity of a model. Henseler and Sarstedt (2013) have shown through a simulation study that the GoF is not a good criterion to evaluate the overall quality of a model, but it can be used to quantify to which extent a model is able to explain different datasets.

[589] Cf. Hair et al. 2011, p. 147; Henseler et al. 2009, pp. 303ff.; Chin 1998b, pp. 316ff..

[590] Cf. Hair et al. 2011, p. 147.
It refers to how many standard deviations a dependent variable changes if the independent variable changes one standard deviation.

[591] Cf. Chin 1998a, p. xiii.

assessed via bootstrapping.[592] The number of samples should be 5,000, and the number of cases should be equal to the sample size (in this case: 351). For the bootstrapping, 351 cases and 5,000 samples were used. Critical t-values for the two-tailed test are 1.65 (significance level = 10 %), 1.96 (significance level = 5 %), and 2.58 (significance level = 1 %).[593] The explained variance is represented by R^2. R^2 of an endogenous latent variable expresses the share of variance that is explained by the related exogenous variables. Values of R^2 range from 0 to 1, and interpretation is analog to traditional regression.[594] The required minimum level depends on the research discipline. In the context of consumer and behavioral studies, R^2 results of 0.2 are already considered high.[595] The effect size f^2 provides information regarding whether any particular exogenous variable has a substantive impact on an endogenous variable. It can be calculated as:

$$f^2 = \frac{R^2_{included} - R^2_{excluded}}{1 - R^2_{included}}$$

$R^2_{included}$ = R^2 of endogenous variable if exogenous variable is used

$R^2_{excluded}$ = R^2 of endogenous variable if exogenous variable is omitted

Effect sizes higher than 0.35, 0.15, and 0.02 can be considered strong, moderate, and weak, respectively.[596]

The last quality criterion assesses how well the structural model is able to predict.[597] In PLS-SEM, the criterion of Stone and Geisser for predictive relevance Q^2 is predominantly applied.[598] Q^2 is computed using the blindfolding procedure in PLS.[599] A value of $Q^2 > 0$ implies that the exogenous variables have predictive relevance for the endogenous construct.[600] Analog to f^2, the relative predictive relevance of the individual exogenous variables on the respective endogenous variables can be assessed by q^2. It is calculated as:

---

[592] See Henseler et al. 2009, pp. 305ff. or Chin 2010 for a detailed description of the bootstrapping procedure.
[593] Cf. Hair et al. 2011, p. 145.
[594] Cf. Chin 1998b, p. 316.
[595] Cf. Hair et al. 2011, p. 147.
[596] Cf. Chin 1998b, p. 317 based on Cohen 1988, p. 355.
[597] Cf. Hair et al. 2011, p. 147.
[598] See Stone 1974 and Geisser 1975.
[599] Cf. Tenenhaus et al. 2005, pp. 174ff.
The blindfolding procedure divides the full data set (across all respondents and items) into groups and then omits one group from the data set at each run. The omitted data points are then predicted with the information from the remaining data. An omission distance OD should be selected so that the number of observations divided by OD is not equal to an integer. Additionally OD should be between five and ten. (Cf. Chin 1998b, p. 318). Finally, an omission distance of 7 was selected.
[600] Cf. Chin 1998b, p. 318; Hair et al. 2011, p. 145.

$$q^2 = \frac{Q^2_{included} - Q^2_{excluded}}{1 - Q^2_{included}}$$

$Q^2_{included} = Q^2$ of endogenous variable if exogenous variable is used

$Q^2_{excluded} = Q^2$ of endogenous variable if exogenous variable is omitted

In line with the evaluation of f^2, the same critical values apply for q^2, i.e., 0.35 indicates a strong, 0.15 a moderate, and 0.02 a weak predictive relevance.[601]

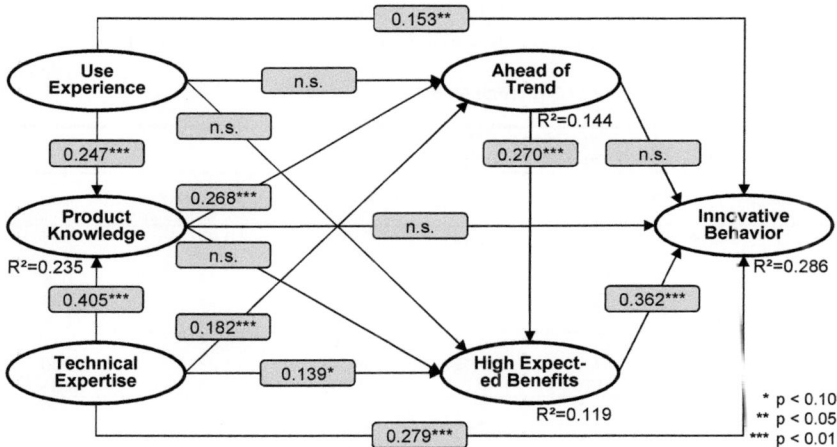

Figure 27: Results of Structural Model 1[602]

Figure 27 and Table 26 show the results of the PLS model estimation. The analysis of path coefficients shows that not all theoretically presumed relationships can be substantiated. Use experience shows no significant or meaningful impact on the two dimensions of lead userness. Use experience does have a highly significant influence on product knowledge ($\gamma_{UE,PK}$ = 0.247, $p < 0.01$) and a slightly low but significant influence on innovative behavior ($\gamma_{UE,IB}$ = 0.153, $p < 0.05$). Technical expertise has a meaningful and significant impact on product knowledge ($\gamma_{TE,PK}$ = 0.405, $p < 0.01$) and considerably influences innovative behavior ($\gamma_{TE,IB}$ = 0.279, $p < 0.01$). The impact of technical expertise on the two dimensions of lead userness is

---

[601] Cf. Henseler et al. 2009, p. 305.
[602] Own illustration, N = 351.
PLS-SEM algorithm settings: path weighting scheme; initial weights: 1.0; abort criterion: $< 10^{-5}$; maximum iterations: 3,000.
Bootstrapping settings: No sign changes; bootstrap samples: 5,000; bootstrap cases: 351

significant, but the size of the impact is rather low ($\gamma_{TE,AoT}$ = 0.182, p < 0.01; $\gamma_{TE,HEB}$ = 0.139, p < 0.05). While the effect size and predictive relevance of technical expertise on being ahead of trend is still weak ($f^2_{TE,AoT}$ = 0.032, $f^2_{TE,AoT}$ = 0.020), the same does not hold true for the relationship to high expected benefits. Product knowledge shows a relevant and significant impact on being ahead of trend ($\gamma_{PK,AoT}$ = 0.268, p < 0.01), but no significant impact on high expected benefits or innovative behavior directly. Being ahead of trend impacts the high expected benefits significantly and with a meaningful size ($\gamma_{AoT,HEB}$ = 0.270, p < 0.01). Interestingly though, the direct effect on innovative behavior is not significant. Finally, the impact of high expected benefits on innovative behavior is strong and highly significant ($\gamma_{HEB,IB}$ = 0.362, p < 0.01).

Table 26: Quality Criteria of Structural Model 1

Endogenous Variable	R^2	Q^2	Exogenous Variable	Path co-efficient γ	T-value	f^2	q^2
Critical Value		$Q^2 > 0$		$\gamma > 0.2$	≥ 1.645 : p < 0.10 ≥ 1.960 : p < 0.05 ≥ 2.580 : p < 0.01	> 0.02 : weak > 0.15 : moderate > 0.35 : strong	> 0.02 : weak > 0.15 : moderate > 0.35 : strong
Innovative Behavior	0.286	0.290	AoT	-0.081	1.562	0.007	0.008
			HEB	0.362	7.336	0.162	0.162
			PK	0.067	1.328	0.004	0.006
			TE	0.279	5.991	0.085	0.089
			UE	0.153	3.051	0.031	0.037
Ahead of Trend	0.144	0.092	PK	0.268	5.349	0.064	0.041
			TE	0.182	4.012	0.032	0.020
			UE	-0.010	0.183	0.000	0.001
High Exp. Benefits	0.119	0.084	AoT	0.270	4.937	0.072	0.050
			PK	0.020	0.353	0.000	0.000
			TE	0.139	2.462	0.017	0.010
			UE	-0.025	0.400	0.001	-0.001
Product Knowledge	0.235	0.140	TE	0.405	7.977	0.212	0.110
			UE	0.247	3.668	0.075	0.043
OD = 7				Cases: 351; Sample: 5,000			

The explained variance of the dependent variable innovative behavior is satisfactory, with R^2_{IB} = 0.286. Although the explained variance of the lead userness dimensions is below the requested value of 0.2 (R^2_{AoT} = 0.144, R^2_{HEB} = 0.119), this does not cause a problem, because both dimensions are often defined as independent variables. Furthermore, it was not expected that use experience, product knowledge, and technical expertise could exhaustively explain the dimensions. All significant and meaningful paths of the model showed effect sizes of considerable levels above the minimum value of 0.02 (see Table 26 above). The effect of high expected benefits on innovative behavior and technical expertise on product knowledge can be considered as moderate ($f^2_{HEB,IB}$ = 0.162; $f^2_{TE,PK}$ = 0.212) while all others are weak.

The Stone-Geisser Q^2 of all endogenous variables are above zero, so that the predictive relevance of the model can be confirmed. The relative predictive relevance of exogenous variables on endogenous variables was confirmed for all relationships with a meaningful and significant path coefficient.

7.2.8 Mediator Analysis for High Expected Benefits

A construct may act as a mediator "*[...] to the extent that it accounts for the relation between the predictor and the criterion.*"[603] As shown in Figure 28 below the effect from construct Y_1 to Y_3 consists of the direct effect $γ_{13}$ and an indirect effect through Y_2 ($γ_{12}$ x $γ_{23}$). If the inclusion of the indirect effect significantly changes the direct effect, then there exists a mediator effect of Y_2 on the relationship between Y_1 and Y_3. Construct Y_2 can be considered a mediator if (a) Y_1 significantly predicts Y_3 (i.e., $γ_{13}$ is relevant and significant), (b) Y_1 significantly predicts Y_2 (i.e., $γ_{12}$ is significant), and (c) Y2 significantly predicts Y_2, controlling for Y_1 (i.e., $γ_{23}$ is significant).[604]

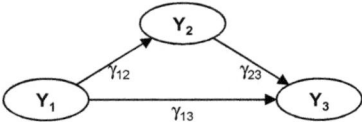

Figure 28: General Mediator Model[605]

The mediator effect of high expected benefits on the relationship of ahead of trend and innovative behavior was tested through the bootstrapping method of Preacher and Hayes (2004) because it does not require distributional assumptions and the bootstrapping function is already implemented in the SmartPLS software. The first condition to establish a mediator effect is that the direct effect from ahead of trend on innovative behavior is significant when the mediator is not yet included in the model. The analysis results in $γ_{AoT,IB}$ = 0.018 with a t-value of 0.333.[606] Since the direct effect is already not significant, no further analysis is required, and there exists no mediation. Nevertheless, there is a significant indirect effect from ahead of trend on innovative behavior of ($γ_{AoT,HEB}$ x $γ_{HEB,IB}$) = 0.098 with p < 0.01. The mediator analysis was also conducted for the SiA and Non-SiA sub-sample, which yielded the same results.

---

[603] Baron & Kenny 1986, p. 1176.
[604] Cf. Preacher & Hayes 2004, p. 717; Baron & Kenny 1986, p. 1176.
[605] According to Hair et al. 2013, p. 220.
[606] Cases: 351; Sample size: 5,000.

7.2.9 Evaluation of Control Variables

The relationships of the main model, including the impact of the control variables (detailed in chapter 7.1.2.4), were analyzed in a separate step. This was mainly done to keep the main model as simple as possible and because the usable sample size, including all control variables, decreases to 256.[607]

Control variables are integrated into a structural model by connecting them with any endogenous variable.[608] The PLS analysis is then conducted as always.

When the control variables *available time, disposable income, education, gender, income, marital status, occupation intensity,* and *origin* are taken into account, the results of the structural model 1 are confirmed.[609] The only notable difference occurs in the causal relationship of product knowledge on innovative behavior, which is significant but still rather low after the control variables are included ($\gamma_{controlled:\ PK,IB} = 0.103$, $p < 0.10$).

The analysis of the effects of the control variables on the endogenous variables shows that most path coefficients are not significant. Those that are, are just under the threshold of $p < 0.10$, or do not show a meaningful strength of $\gamma > 0.2$. The only meaningful and highly significant relationship is from available time on product knowledge ($\gamma_{Time,PK} = 0.220$, $p < 0.10$). Further significant effects among control variables include the following:

1. Low negative impact of *education* on *product knowledge* ($\gamma_{Edu.,PK} = -0.108$, $p < 0.10$),
2. Low positive impact of *education* on *high expected benefits* ($\gamma_{Edu.,HEB} = 0.148$, $p < 0.05$),
3. Low negative impact of *gender* on *product knowledge* ($\gamma_{Gender,PK} = -0.107$, $p < 0.05$),
4. Low positive impact of *income* on *product knowledge* ($\gamma_{Income,PK} = 0.137$, $p < 0.05$),
5. Moderate positive impact of *income* on *being ahead of trend* ($\gamma_{Income,AoT} = 0.184$, $p < 0.01$),
6. Low positive impact of *marital status* on *high expected benefits* ($\gamma_{Marital\ Status,HEB} = 0.162$, $p < 0.05$),

---

[607] The sharp reduction in complete answers is mainly due to the questions on income (52 missing answers), disposable income (49 missing answers), and available time (22 missing answers), which some respondents chose not to answer.
[608] Cf. Kock et al. 2008, p. 188.
[609] The strength of the path coefficient as well as t-values only change slightly. For the detailed values of all path coefficients and t-values, please refer to Appendix 5.

7. Low positive impact of *occupation intensity* on *high expected benefits* ($\gamma_{Occupation,HEB} = 0.128$, $p < 0.10$),
8. Low positive impact of *occupation intensity* on *innovative behavior* ($\gamma_{Occupation,IB} = 0.120$, $p < 0.10$),
9. Very low positive impact of *origin* on *innovative behavior* ($\gamma_{Origin,IB} = 0.097$, $p < 0.10$).

7.2.10 Interaction Effect of Age on Structural Model

The simple and moderator effect of age (both chronological and cognitive age) on the latent variables and the relationships between them in structural model 1 should be analyzed. While the simple effects can be easily assessed by adding a construct age to the structural model, the moderator effect requires a more sophisticated approach. A moderator is defined as "*[...] a qualitative (e.g., sex, race, class) or quantitative (e.g., level of reward) variable that affects the direction and/or strength of the relation between an independent or predictor variable and a dependent or criterion variable.*"[610] Moderator values can either be metric (e.g., chronological age, income) or categorical (e.g., gender, occupation). According to the type of scale, there are two common approaches for the estimation of moderator effects: the product-term approach and the group comparison.[611] The product-term approach is mostly applied for metric moderators. In the product-term approach, a so-called interaction term is included into the structural model, which is defined as the product of all indicators of the exogenous and the moderator variable. The path coefficient from the interaction term to the endogenous variable can be interpreted as the moderator effect of the moderator on the path coefficient from the exogenous on the endogenous variable (see also Figure 29 below).[612]

If the measurement model includes formative measurements, the product-term approach, which is based on indicators, leads to biased results, and a two-stage calculation approach is proposed.[613] During this approach, the PLS algorithm is run without the interaction term, and the latent variable scores are extracted. From the extracted latent variable scores, the interaction term is calculated and then used as an independent value during the second run of the PLS algorithm.[614] A comparison of available approaches for the analysis of interaction effects by Henseler and Chin

---

[610] Cf. Baron & Kenny 1986, p. 1174.
[611] Cf. Henseler & Fassott 2010, p. 718.
[612] Cf. Baron & Kenny 1986, p. 1174; Henseler & Fassott 2010, pp. 718f.
[613] Cf. Chin et al. 2003, pp. Appendix D; Henseler & Fassott 2010, p. 724.
[614] Cf. Henseler & Fassott 2010, pp. 724f.; Chin et al. 2003, pp. Appendix D.

(2010) resulted in the recommendation of the two-stage approach, especially *"when a researcher is mainly interested in the significance of an interaction effect"*[615].

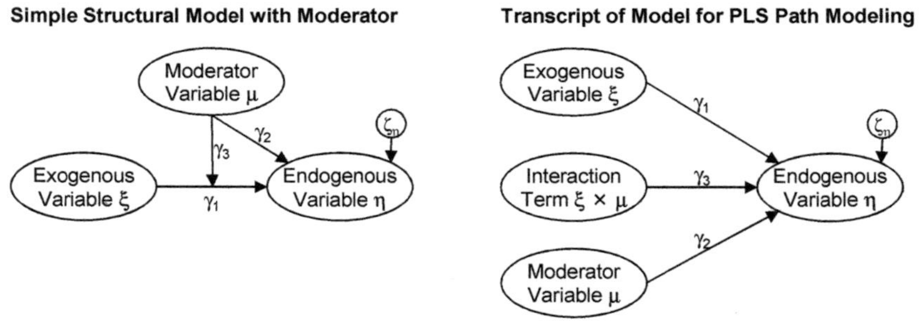

Simple Structural Model with Moderator   **Transcript of Model for PLS Path Modeling**

Figure 29: Transcription of Structural Model with Moderator for PLS Path Modeling[616]

Table 27 below summarizes the result of the PLS estimation of interaction model 1 (with chronological age as moderator) and interaction model 2 (with cognitive age as moderator). The simple effect of chronological age on the latent variables is significant and highly positive for use experience ($\gamma_{Chron.\ Age,UE}$ = 0.551, $p < 0.01$). This is not surprising, as a high age is a prerequisite for a long duration of existing experience. The simple effects of chronological age on technical expertise and being ahead of trend are negative on a low level and are significant. ($\gamma_{Chron.\ Age,TE}$ = -0.176, $p < 0.01$; $\gamma_{Chron.\ Age,AoT}$ = 0.109, $p < 0.05$). The only significant moderator effect of chronological age affects the relationship of use experience and innovative behavior negatively ($\gamma_{Chron.\ Age*UE,IB}$ = -0.104, $p < 0.10$). The interpretation of this moderator effect is as follows: The simple effect of use experience on innovative behavior is 0.226, which means that a change of one standard deviation in use experience results in a change of 0.226 standard deviations in innovative behavior, under the condition that the value of the standardized moderator chronological age is equal to zero. A change of one standard deviation in chronological age decreases the path coefficient from use experience to innovative behavior by -0.104 to 0.122. No other relationships in interaction model 1 are affected by the moderator chronological age.

---

[615] Henseler & Chin 2010, p. 105.
[616] Own illustration according to Baron & Kenny 1986, p. 1174 and Henseler & Fassott 2010, pp. 717 & 719.

Table 27: Results of Interaction Model 1 and 2

	Exogenous Variable	Endogenous Variable	Interaction Model 1 Moderator: Chronol. Age		Interaction Model 2: Moderator: Cognitive Age	
			Path Coefficient	T-Value	Path Coefficient	T-Value
Simple Effects	Use Experience	PK	0.234	4.349	0.281	5.069
		AoT	0.023	0.345	0.054	0.085
		HEB	0.039	0.610	0.008	0.122
		IB	0.226	3.820	0.233	3.950
	Product Knowledge	AoT	0.266	4.975	0.277	5.394
		HEB	0.039	0.663	0.041	0.735
		IB	0.070	1.345	0.076	1.487
	Technical Expertise	PK	0.412	8.448	0.416	8.779
		AoT	0.163	3.282	0.158	3.207
		HEB	0.113	1.832	0.128	2.030
		IB	0.259	4.958	0.285	5.402
	Ahead of Trend	HEB	0.287	4.935	0.269	4.968
		IB	-0.072	1.365	-0.088	1.682
	High Expected Benefits	IB	0.361	7.000	0.361	7.189
	Moderator	UE	0.551	14.459	0.531	12.475
		PK	0.018	0.309	-0.018	0.323
		TE	-0.176	3.196	-0.247	4.507
		AoT	-0.109	1.756	-0.128	2.331
		HEB	-0.035	0.584	-0.053	0.871
		IB	-0.062	1.017	-0.089	1.461
	Interaction Term					
Moderator Effect	Moderator*UE	PK	-0.059	0.779	-0.018	0.286
		AoT	0.038	0.616	0.010	0.198
		HEB	-0.063	1.312	-0.086	1.794
		IB	-0.104	1.921	-0.088	1.679
	Moderator*PK	AoT	-0.022	0.409	-0.053	1.237
		HEB	-0.026	0.524	-0.069	1.486
		IB	-0.030	0.677	-0.071	1.558
	Moderator*TE	PK	0.038	0.488	-0.027	0.429
		AoT	-0.022	0.401	-0.030	0.635
		HEB	-0.026	0.520	-0.022	0.425
		IB	-0.030	0.671	-0.065	1.429
	Moderator*AoT	HEB	0.046	0.955	0.027	0.559
		IB	0.017	0.365	0.007	0.156
	Moderator*HEB	IB	-0.042	0.884	-0.044	0.932
	Critical T-Values: ≥ 1.645 : p < 0.10 ≥ 1.960 : p < 0.05 ≥ 2.580 : p < 0.01		Cases: 333 Samples: 5,000		Cases: 351 Samples: 5,000	

The results of interaction model 2 with cognitive age as the moderator are similar, although not identical. The simple effect of cognitive age on use experience is strong and highly significant ($\gamma_{Cogni.\ Age,UE}$ = 0.531, p < 0.01). There are also negative significant effects on being ahead of trend ($\gamma_{Cogni.\ Age,AoT}$ = -0.128, p < 0.05) and technical experience ($\gamma_{Cogni.\ Age,TE}$ = -0.247, p < 0.01). Like interaction model 1, the only significant moderating influence of cognitive age is on the relationship of use experience and innovative behavior ($\gamma_{Cogni.\ Age*UE,IB}$ = -0.088, p < 0.10), but the effect is slightly weaker. Additionally, cognitive age moderates the relationship of use experience on high expected benefits, which itself is not significant.

Henseler and Fassott (2010) also suggest applying the product-term approach for categorical moderators after transforming them into dummy variables. This approach provides good results on whether a moderator effect is significant or not, but the size of the group differences itself remains unclear. The product-term approach also provides results for the full range (from minimum to maximum) of the moderator variable. For the comparison of age groups, however, the difference between specific age groups is relevant. Therefore, the multi-group analysis (MGA) was applied to compare the SiA group with the Non-SiA group, in accordance with suggestions.[617] During an MGA, the full data set is divided into separate data sets, with the moderator as a grouping variable. The structural model is then estimated for all sets of data, and the sizes of values of path coefficients are compared among groups. The moderator effect is then calculated as the differences between path coefficients:

$$\gamma_{Moderator} = \gamma^{(1)} - \gamma^{(2)}$$

To test whether this moderator effect is significant, Henseler, Ringle, and Sinkovics (2009) proposed the PLS-MGA approach, which considers the observed distribution of the bootstrapping results.[618] This approach does not require normal distribution and is similar to the Mann-Whitney-U-test.[619] Comparison with other procedures to test for group differences has shown that the PLS-MGA approach is more conservative.[620] The probability can be calculated as[621]:

$$P\left(\gamma^{(1)} > \gamma^{(2)} | \beta^{(1)} \leq \beta^{(2)}\right) = 1 - \sum_{\forall j, i} \frac{\Theta\left(2\bar{\gamma}^{(1)} - \gamma_j^{(1)} - 2\bar{\gamma}^{(2)} + \gamma_i^{(2)}\right)}{J^2}$$

$\gamma^{(1)}; \gamma^{(2)}$ = Parameter estimates of group 1 and group 2
$\beta^{(1)}; \beta^{(2)}$ = True parameters of group 1 and group 2
$\gamma_j^{(1)}; \gamma_i^{(2)}$ = Bootstrap parameter estimates
$\bar{\gamma}^{(1)}; \bar{\gamma}^{(2)}$ = Means of focal parameters over bootstrap samples
Θ = Unit step function [$\Theta(>0)=1$, otherwise 0]
J = Number of bootstrap samples

---

[617] Cf. Rigdon et al. 1998, p. 1.
[618] Cf. Henseler et al. 2009, p. 309.
[619] See Mann & Whitney 1947.
[620] Cf. Sarstedt et al. 2011, p. 213. Comparison of parametric approach, permutation-based approach, PLS-MGA approach, and nonparametric confidence set approach showed that the latter two identify fewer significant group differences and can therefore be regarded as more conservative.
[621] Cf. Henseler et al. 2009, p. 309.

According to the definition of the SiA segment (see chapter 2.2), all respondents of at least 55 years of age were assigned to the SiA group, and all respondents below 55 years were assigned to the Non-SiA group. The SiA group contained 110 respondents, and the Non-SiA group contained 223.[622] As was shown in chapter 0, most older people regard themselves as younger than they actually are, and the difference between chronological and cognitive age varies greatly. This makes defining a precise grouping value even more difficult. Therefore, the upper third of cognitive age values was assigned to the High Cognitive Age group and the lower third to the Low Cognitive Age group, respectively. The middle third was not assigned to any group.[623] The High Cognitive Age group consisted of 128 respondents with a minimum cognitive age of 47.0 years and an average of 55.3 years. The Low Cognitive Age group consisted of 119 respondents with a maximum cognitive age of 38.3 years and an average of 33.5 years.[624]

Table 28: Results of Multi-Group Analysis Regarding Age

Exogenous Variable	Endogenous Variable	Silver Age Group γ_{SiA}	Non-Silver Age Group $\gamma_{Non\,SiA}$	Group Differences Moderator	p-value	High Cognitive Age Group γ_{HighCA}	Low Cognitive Age Group γ_{LowCA}	Group Differences Moderator	p-value
Use Experience	PK	$0.119^{n.s.}$	0.250^{***}	-0.131	0.165	0.119^{*}	0.217^{*}	-0.023	0.438
	AoT	$0.054^{n.s.}$	$0.014^{n.s.}$	0.040	0.623	$0.086^{n.s.}$	$0.055^{n.s.}$	0.032	0.581
	HEB	$-0.033^{n.s.}$	$0.019^{n.s.}$	-0.052	0.344	$-0.056^{n.s.}$	$0.017^{n.s.}$	-0.073	0.325
	IB	$0.027^{n.s.}$	0.290^{***}	-0.263	0.013	$0.067^{n.s.}$	$0.101^{n.s.}$	-0.034	0.387
Product Knowledge	AoT	0.235^{***}	0.291^{***}	-0.056	0.297	0.227^{***}	0.240^{***}	-0.012	0.450
	HEB	-0.159^{*}	0.139^{*}	-0.297	0.005	-0.118^{*}	$0.034^{n.s.}$	-0.152	0.129
	IB	$0.046^{n.s.}$	$0.047^{n.s.}$	-0.001	0.498	$0.027^{n.s.}$	0.143^{*}	-0.116	0.165
Technical Expertise	PK	0.441^{***}	0.419^{***}	0.023	0.598	0.388^{***}	0.445^{***}	-0.056	0.317
	AoT	0.200^{***}	0.136^{***}	0.064	0.732	0.183^{**}	0.169^{**}	0.014	0.554
	HEB	0.169^{*}	$0.083^{n.s.}$	0.086	0.761	0.206^{**}	0.242^{**}	-0.036	0.389
	IB	0.327^{***}	0.212^{***}	0.115	0.860	0.294^{***}	0.210^{***}	0.084	0.764
Ahead of Trend	HEB	0.411^{***}	0.217^{***}	0.194	0.945	0.329^{***}	0.194^{***}	0.135	0.843
	IB	$0.033^{n.s.}$	-0.109^{*}	0.142	0.883	$0.024^{n.s.}$	$-0.049^{n.s.}$	0.090	0.766
High Exp. Benefits	IB	0.192^{*}	0.417^{***}	-0.225	0.027	0.213^{**}	0.447^{***}	-0.234	0.028

* p < 0.10
** p < 0.05
*** p < 0.01

Cases: 110 Cases: 223
Samples: 5,000

Cases: 128 Cases: 119
Samples: 5,000

---

[622] 18 respondents could not be assigned to any of these groups because they did not indicate their chronological age.
[623] Cf. Henseler & Fassott 2010, p. 720.
[624] The High Cognitive Age group contained 23 respondents, who were assigned to the Non-SiA group, and 7 respondents, who were not assigned to any SiA group, because they did not indicate their age. The Low Cognitive Age group contained 6 previously not assigned respondents, due to a missing chronological age, and one respondent who was assigned to the SiA group.

Before group-related difference can be compared, one has to test for measurement invariance to ensure that the grouping variable only impacts differences in the path coefficients and not in the measurement model.[625] The guidelines of Chin and Dibbern (2010) were followed to establish measurement invariance between the groups formed based on chronological and cognitive age.[626]

Table 28 above shows the results of the MGA for chronological and cognitive age. Differences between groups can be considered significant if only one of the two path coefficients is significant or if the p-value is above 0.9 or below 0.1. According to this criterion, seven path coefficients are significantly different between the SiA and the Non-SiA group:

- Use experience has a highly significant and meaningful influence on product knowledge and the innovative behavior for the Non-SiA group, while this influence completely diminishes in the SiA group.
- While product knowledge has a positive impact on high expected benefits in the younger group, the sign changes in the SiA group.
- Technical expertise only has an impact on high expected benefits in the SiA group.
- The relationship among the dimensions of lead userness is much stronger in the SiA group.
- Being ahead of trend only had a small negative impact on innovative behavior in the Non-SiA group.
- The impact of high expected benefits on the innovative behavior is stronger for the Non-SiA group.

The explained variance R^2 for some latent variables also differed among age groups. While levels for product knowledge were comparable ($R^2_{SiA:PK} = 0.235$, $R^2_{NonSiA:PK} = 0.259$) and ahead of trend ($R^2_{SiA:AoT} = 0.150$, $R^2_{NonSiA:AoT} = 0.141$), differences for high expected benefits ($R^2_{SiA:HEB} = 0.195$, $R^2_{NonSiA:HEB} = 0.118$) and innovative behavior ($R^2_{SiA:IB} = 0.209$, $R^2_{NonSiA:IB} = 0.367$) were quite obvious. The general model to predict innovative behavior fits much better for the Non-SiA group than for the SiA group.

---

[625] Cf. Sarstedt et al. 2011, p. 199.
[626] For detailed results see Appendix 3 and Appendix 4 below. Measurement invariance was established for all items except in the case of UE [1]. The outer weight of UE [1] and, therefore, its impact is rather low anyway. The group differences based on a structural model excluding UE [1] were also calculated. The results differed only marginally, so UE [1] was kept in the measurement models of both groups. In the Low Cognitive Age group, the outer loading of PK [1] was 0.663, which is just below the suggested value of 0.7. Items only have to be deleted if their loading is below 0.4 or if their deletion significantly improves composite reliability and average variance extracted, so it was retained for better comparability. Cf. Hair et al. 2013, p. 104.

The comparison of groups with high and low cognitive age shows only two significant differences: the impact of product knowledge on innovative behavior is only significant for the Non-SiA group, and the impact of high expected benefits on innovative behavior is much stronger for the younger group. This confirms the result of the MGA based on chronological age.

The same PLS-MGA was also conducted for all sub-dimensions of cognitive age with the respective median age per sub-dimensions used as a separating value between the groups. Across all sub-dimensions, use experience only has an impact on innovative behavior for the younger group. The DO age group did not show any other relevant differences. There were two more differences between the groups according to the INTEREST age dimension: product knowledge had a positive impact on innovative behavior for respondents with a low interest age and technical expertise impacted high expected benefits only for the older group. The FEEL age groups differed on the relationships between technical expertise and being ahead of trend and between high expected benefits and innovative behavior. The first relationship was only significant for the group feeling older, while the second relationship was stronger for the group that felt younger. The two LOOK age groups showed the most differences. In addition to the difference in the relationship between high expected benefits and innovative behavior (significantly stronger for younger looking respondents), younger looking respondents showed a significantly lower impact of technical expertise on being ahead of trend and innovative behavior, as well as a much stronger positive and more significant impact of use experience on product knowledge. Lastly, the negative impact of being ahead of trend on innovative behavior was only detectable for younger looking respondents. Detailed results of all analyses can be found in Appendix 5.

7.2.11 Testing for Non-Linear Effects from Use Experience

The PLS-MGA analyses in the previous chapter have shown use experience has a stronger influence on innovative behavior for younger people than for older people.[627] Correlation coefficients between chronological age and the two components of use experience are low but highly significant ($r_{\text{Chronological Age,UE [1]}} = 0.330$, $p < 0.01$; $r_{\text{Chronological Age,UE [2]}} = 0.443$, $p < 0.01$).[628] This correlation coefficient is higher for

---

[627] Path coefficient in Non-SiA group ($\gamma_{UE,IB}^{NonSiA}=0.290$, p<0.01) was much higher and significant compared to path coefficient in SiA group ($\gamma_{UE,IB}^{SiA}=0.027$, n.s.). Moderator analysis with chronological age as moderator showed comparable results. The moderating influence of chronological age on the relationship of use experience and innovative behavior was significant with $\gamma_{\text{Chronological Age*UE,IB}} = -0.104$, $p < 0.10$.

[628] Correlation coefficient calculated according to Spearman's rho, due to non-parametric distribution of variables.

duration of use experience (UE [2]) than for frequency (UE [1]). This is not surprising because a person cannot accumulate more years of use experience than his or her own actual age. This suggests that the differing impacts of use experience on innovative behavior between the age groups might not be caused by chronological age. Rather, it might actually be a sign for a negative non-linear effect of use experience.

PLS is already equipped to calculate non-linear effects because non-linear effects can be interpreted as a self-moderation of a construct on their own relationships.[629] Therefore, the approaches to moderator analysis can also be applied to evaluate non-linear effects.[630] To calculate the non-linear effect, Henseler et al. (2012a) suggest applying the two-stage approach detailed in chapter 7.2.10 above. The results showed that the only significant non-linear effect exists between use experience and innovative behavior. Although the effect is weak ($f^2_{UE^2,IB} = 0.020$), it is nevertheless of a notable size and significant ($\gamma_{UE^2,IB} = -0.137$, $p < 0.05$). The negative effect size means that use experience has a diminishing impact on innovative behavior, i.e., the higher the use experience, the weaker the positive impact on innovative behavior.[631] This is an indicator for the existence of functional fixedness,[632] although some authors argue that functional fixedness should not be an issue for lead users.[633]

7.2.12 Characterization of Silver Market User Innovators and Non-Innovators

For researchers interested in conducting behavioral studies (especially on user innovation) among older users, it would be of great value to know if and how user innovators differ from non-innovators in easy to measure demographical characteristics. The same applies to manufacturers of age-based products who are searching for user innovators in their fields. For this reason, the demographic characteristics of innovators and non-innovators of at least 55 years of age were compared. The groups were tested for significant differences in group averages using the Mann-Whitney U-test.[634] Table 29 below provides an overview of the results.

---

[629] Cf. Rigdon et al. 2010, pp. 262f.
[630] Cf. Henseler & Chin 2010, p. 107; Wold 1982.
[631] The simple linear effect of use experience on innovative behavior in this case is $\gamma_{UE,IB} = 0.226$, $p < 0.01$.
[632] See Adamson 1952.
[633] Cf. Lüthje & Herstatt 2004, p. 557.
[634] See Mann & Whitney 1947.

Surprisingly, the groups do not differ in chronological age.[635] Instead, innovators have a significantly lower cognitive age (3.6 years), which is especially evident in the *FEEL* and *DO* age (4.9 years lower). Concurrently, the age difference of innovators (-10.0 years) is larger than for non-innovators (-7.2 years). The only cognitive age dimension that did not show significant differences is the *INTEREST* age.

Table 29: Comparison of Characteristics of Innovators and Non-Innovators in the Silver Market Segment

Category	Characteristic	Innovators	Non-Innovators	Δ	p-value[a)]
Age	Average	62.9 y	63.7 y	-0.8 y	n.s.
Cognitive Age	Overall *(average)*	52.9 y	56.5 y	-3.6 y	< 0.05
	FEEL age *(Ø)*	51.7 y	56.6 y	-4.9 y	< 0.05
	LOOK age *(Ø)*	56.0 y	58.5 y	-2.5 y	< 0.10
	DO age *(Ø)*	50.4 y	55.3 y	-4.9 y	< 0.05
	INTEREST age *(Ø)*	53.7 y	55.5 y	-1.8 y	n.s.
	Age difference *(Ø)*	-10.0 y	-7.2 y	-2.8 y	< 0.05
Gender	Male	93.0 %	80.6 %	12.4 %	< 0.10
	Female	7.0 %	19.4 %	-12.4 %	
Income	Monthly household inc. *(Ø)*	3,167 EUR	3,074 EUR	93 EUR	n.s.
	Disposable income *(Ø)*	32.2 %	33.2 %	-1.0 %	n.s.
Available Time	Average	7.0 h	7.2 h	-0.2 h	n.s.
Origin	Federal Republic of Germany	62.6 %	67.3 %	-4.7 %	n.s.
Education	Secondary school	4.9 %	12.1 %	-7.2 %	
	Intermediate school	9.8 %	16.7 %	-6.9 %	
	High school	17.1 %	3.0 %	14.1 %	n/a
	Apprenticeship	39.0 %	48.5 %	-9.5 %	
	University degree	29.3 %	19.7 %	9.6 %	
Occupation Intensity	Full-time	48.8 %	22.7 %	26.1 %	
	Part-time	4.7 %	10.6 %	-5.9 %	n/a
	Unemployed	0.0 %	1.5 %	-1.5 %	
	Retired	46.5 %	65.2 %	-18.7 %	
Marital Status	Single	0.0 %	3.0 %	-3.0 %	
	In a partnership	9.3 %	3.0 %	6.3 %	
	Married	86.0 %	89.6 %	-3.6 %	n/a
	Divorced	2.3 %	3.0 %	-0.7 %	
	Widowed	2.3 %	1.5 %	0.8 %	
Use Experience	Intensity (UE [1]) *(Ø)*	48.2 d/y	46.3 d/y	1.9 d/y	n.s.
	Duration (UE [2]) *(Ø)*	29.7 y	25.7 y	4.0 y	n.s.
Construct Scores	Product Knowledge	4.2	3.8	0.4	< 0.05
	Technical Expertise	4.2	3.2	1.0	< 0.01
	Ahead of Trend	2.6	2.1	0.5	< 0.05
	High Expected Benefits	3.4	2.4	1.0	< 0.01
		N = 43	N = 67		

a) Calculated with Mann-Whitney U-test

In line with previous research, innovators are predominantly male, well educated, and employed full-time.[636] Although the average age was comparable, non-innovators are

---

[635] Previous studies on user innovators among younger age groups typically found that user innovators tend to be younger. Cf. Im et al. 2003, p. 63; Steenkamp et al. 1999, p. 65; Flowers et al. 2010, p. 17.

[636] Cf. Hippel et al. 2011, p. 31; Flowers et al. 2010, p. 5.

more likely to be retired (65 % versus 47 %), indicating that being able to work and staying active has a positive impact on innovative behavior. Income levels, available time, origin (only applicable for German users), and marital status were comparable. The numbers for use experience were generally very high, indicating that older camping tourists have more time available to exercise camping and do so since at least two decades (on average). Levels of the intensity and duration of use experience did not show significant differences in the SiMa segment, which is in line with the finding that use experience is not a good indicator of the innovative behavior of older users (see previous chapters 7.2.10 and 7.2.11). Scores for the theoretical constructs were significantly higher for innovators across all products. Innovators assessed their technical expertise, in particular, as being better and expected greater benefits from a product innovation.

7.2.13 Evaluation of Hypotheses

The analyses above have shown that the measurement model and structural model are valid and reliable. The testing of hypotheses is therefore legitimate, and the results are meaningful. Table 30 below summarizes the results of the hypotheses testing.

The first section on the determinants of innovative behavior consisted of fifteen hypotheses. Ten of the hypotheses are supported. The remaining five hypotheses were rejected due to insufficient significance levels. Four of the five hypotheses also did not show a sufficiently strong effect.

The second section on the moderating influence of chronological age consisted of twelve hypotheses. Six of the hypotheses were supported[637] and six were rejected. In four of the six cases, the hypothesis was rejected because there was no significant effect. In the remaining two cases, the effect was significant but was of an opposite direction or age was expected to not have a moderating influence.

---

[637] In the case of H12, the difference of the path coefficients was 0.142 with a p-value of 0.883. The required p-value to be considered significant was defined at $p \geq 0.9$. Although the difference was slightly not significant, the hypotheses were still supported because the path coefficient was only significant in the Non-SiA group and not in the SiA group. Therefore, the difference between the groups is by definition significant.

Table 30: Evaluation of Hypotheses

	Hypothesis	p-value	Direction of effect[a]	Support of hypothesis
Hypotheses on the Determinants of Innovative Behavior				
H1a	UE is positively related to being AoT.	n.s.	O	No
H1b	UE is positively related to HEB.	n.s.	O	No
H1c	UE is positively related to innovative behavior.	< 0.05	+	Yes
H1d	UE is positively related to PK.	< 0.01	+	Yes
H2a	PK is positively related to being AoT.	< 0.01	+	Yes
H2b	PK is positively related to HEB.	n.s.	+	No
H2c	PK is positively related to innovative behavior.	n.s.	O	No
H3a	TE is positively related to being AoT.	< 0.01	+	Yes
H3b	TE is positively related to HEB.	< 0.10	+	Yes
H3c	TE is positively related to innovative behavior.	< 0.01	+	Yes
H3d	TE is positively related to PK.	< 0.01	+	Yes
H4	The lead user component AoT is strongly positively related to the lead user component HEB.	< 0.01	+	Yes
H5	Being AoT is positively related with innovative behavior.	n.s.	O	No
H6	HEB are positively related with innovative behavior.	< 0.01	+	Yes
H7	HEB do not mediate the relationship between AoT and innovative behavior.	n.s.	O	Yes
Hypotheses on the Moderating Influence of Chronological Age[b]				
H8a	Age negatively moderates the impact of UE on AoT.	n.s.	O	No
H8b	Age negatively moderates the impact of UE on HEB.	n.s.	O	No
H8c	Age negatively moderates the impact of UE on IB.	< 0.05	−	Yes
H9a	Age negatively moderates the impact of PK on AoT.	n.s.	O	No
H9b	Age negatively moderates the impact of PK on HEB.	< 0.01	−	Yes
H9c	Age negatively moderates the impact of PK on IB.	n.s.	O	No
H10a	Age does not moderate the impact of TE on being AoT.	n.s.	O	Yes
H10b	Age does not moderate the impact of TE on HEB.	n.s.	O	Yes
H10c	Age does not moderate the impact of TE on IB.	n.s.	+	No
H11	Age positively moderates the impact of AoT on HEB.	< 0.05	+	Yes
H12	Age positively moderates the impact of AoT on IB.	n.s.	+	Yes
H13	Age positively moderates the impact of HEB on IB.	< 0.05	−	No

a) Only differences with an absolute difference to 0 of at least 0.1 were considered identifiable.
b) Evaluation based on the multi-group-analysis

7.3 Findings Regarding Silver Market User Innovations and Related Processes

7.3.1 Descriptive Analysis of Survey Results

In this chapter, the differences between older and younger user innovators regarding the innovation process and its outcomes are analyzed. The analysis is divided into three parts. Firstly, process qualities like furthest development stage, required time, and support by other users are examined in chapter 7.3.1.1. Secondly, the type of the innovation (according to the self-classifications of the users) is the focus of chapter 7.3.1.2. Finally, the qualities of the innovation will be detailed in chapter 7.3.1.3.

The separation of respondents into SiA and Non-SiA groups for the following analyses is not based on current chronological age but on the true age at the time of the innovation. This reduces the size of the SiA group to 27 and the Non-SiA group to 120.[638] Due to the very different sizes of the groups and the small sample size of SiMa user innovators, inferential statistics could not be applied.

7.3.1.1 Findings Regarding Process Qualities

Respondents to the survey were asked questions regarding their latest innovation. The questions on the innovation process inquired about the development stage of the innovation and the time required, whether the innovator had support during the ideation and realization phase and how frequently the user becomes innovative.

Of the total sample, 25 % had a possible solution in mind. Only a few innovators stopped at the stage of concept drawings (10 %). A majority (55 %) built a prototype for personal use. Around 10 % of the innovators have distributed their innovation to other users. 3 % of innovations are already commercialized (see Figure 30 below). These results are in line with previous research, with the only difference being that fewer camping tourists stop at making concept drawings and actually proceed to build a working prototype.[639] The differentiation according to age groups uncovers a major difference: SiA user innovators stop much more often at the idea stage. Compared to 78 % of Non- SiAs, only 63 % of SiAs make concept drawings. This loss of ideas results in fewer prototypes: 58 % of Non-SiAs build a working prototype, while only 41 % of SiAs do the same. In the end, 52 % of SiA user innovators and 69 % of Non-SiA user innovators create a usable product.

One reason for this drop might be that SiA user innovators improve their equipment more frequently. 30 % reported that they improve *(almost) every* major piece of camping equipment they own, compared to only 6 % of Non-SiA user innovators. Generally, almost all user innovators are "serial innovators", with 98 % stating that they have improved their equipment more than once. 33 % of the total sample reported that they improve their equipment at least most of the time. This share was even higher for SiA user innovators, at 48 % (see Appendix 7, Figure 34 for more details). Having more ideas puts pressure on one's resources and requires prioritization. Some innovators might, therefore, abandon some of their ideas at an early stage in order to pursue more promising ones.

---

[638] The true age by the time of the innovation of ten of the 157 innovators could not be determined because they did not share their current chronological age. These ten respondents were not included in the following analyses.

[639] Cf. Lüthje et al. 2005, p. 957. In their study 27.0 % of user innovators reported to have a possible solution in mind, 23.4 % had made concept descriptions / drawings, 40.5 % have built a prototype for themselves, and 9.1 % reported that their idea is already been used by others.

How far have you developed your idea to date?

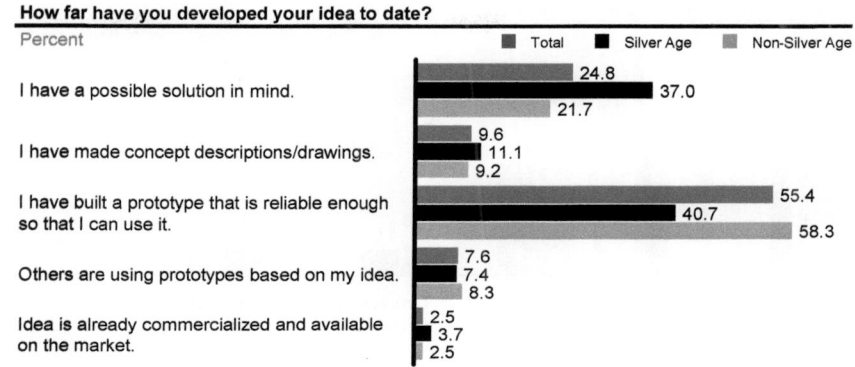

Figure 30: Furthest Development Stage of Innovations[640]

Half of the innovators require less than a month to develop their ideas. In general, SiA user innovators spend a little more time on the development of their ideas, which also correlates with them having more time at their disposal (see Appendix 7 for details).[641]

The community notion of camping and caravanning is also evident during the process of developing a new idea and transforming it into a working product. Almost none of the respondents claimed that they developed the original idea by themselves only (only 2 % of user innovators stated that this was the case). 98 % said that the idea was developed jointly with others. Since camping activities are almost always conducted with one's spouse, family, or friends, it is no surprise that the realization of the shortcomings of existing products and potential ideas for improvement are developed collectively. The high share of collaboration declines during the realization of the ideas, with 34 % of innovators requiring support from others. Main differences between age groups with regard to cooperation were not discovered.[642]

---

[640] Own illustration, N = 157. Statement of the furthest achieved development stage of the indicated innovation. i.e., innovators who stated to have built a working prototype have already achieved (and passed) the stages of solution development and concept description / drawing. Classification into SiA and Non-SiA group according to adjusted innovation age, i.e., the true chronological age at the time of the innovation.

[641] SiAs stated to have 7.5 hours per day disposable time, compared to 4.4 hours per day for Non-SiAs.

[642] 35 % of Non-SiA user innovators required support during realization compared to only 31 % of SiA user innovators, but this difference is not statistically significant.

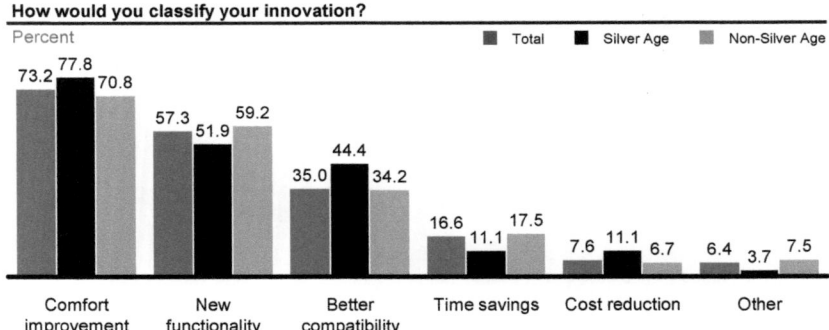

Figure 31: User Self-Classifications of Innovations[643]

7.3.1.2 Findings Regarding Innovation Type

Respondents were asked to classify their innovation according to its type of the improvement, i.e., increased comfort, new functionality, cost or time savings, better compatibility or other reasons. Multiple responses were allowed.

Comfort improvements (73 %), new functionalities (57 %), and a better compatibility (35 %) were the most common types of innovations. Savings in time (17 %) or cost (8 %) showed to have only a low importance (see Figure 31 above).[644] Ideas from SiA user innovators focus more on comfort improvements (+ 7 %) and better compatibility (+ 10 %) than those from Non-SiA user innovators. The difference in the type *better compatibility* is demonstrated in the innovation descriptions. Younger camping tourists want to improve compatibility with their sporting equipment, while older ones want to improve the mounting and innovate to overcome the absence of replacement parts for their aged equipment. Four sample descriptions of better compatibility innovations from the sample illustrate this:

- Absence of replacement parts for aged equipment: "*Refurbished my drawbar kitchen and added some improvements. There are no spare parts available, because year of manufacture is 1985.*" (Male, age at innovation: 65 years)
- Improved mounting of parts in the camper and general safety: "*Improved mounting of drawers so they don't slide out accidentally, developed click-mounting for all camping furniture, added electric illumination in the rear garage.*" (Male, age at innovation: 61 years)

---

[643] Own illustration, N = 157. Classification to SiA and Non-SiA group according to adjusted innovation age, i.e., the true chronological age at the time of the innovation.
[644] In the category *other* especially "security" and "improved stability" were mentioned more than once.

- Added mounting options for sporting equipment: *"Strengthening of the roof so it could support boat transports"* (Male, age at innovation: 41 years); *"Built a carrier that would support my bikes, scooter and kayaks including TÜV certification."* (Male, age at innovation: 52 years)

7.3.1.3 Findings Regarding Innovation Qualities

Users were asked to rate the quality of their own innovation according to the categories newness of the idea, technical quality of the solution, creativity of the idea, benefit to other users (today and in the future), and the sales potential (today and in the future). Self-ratings of user innovations might contain a positive bias because users might overrate the quality and appeal of their own idea. Lüthje, Herstatt, and Hippel (2005) could show for the case of user innovations in mountain biking that although users evaluated their own innovations slightly more favorably than independent experts, the difference was statistically not significant.[645] Therefore, innovators' self-ratings were also relied upon to analyze the sample of this study (see Table 31 below).

User innovators in the sample rate the creativity and benefits to others as favorable, with average ratings above the mid-point of the scale at three. Newness and technical quality are evaluated slightly below the mid-point, but all ratings are in line with previous research.[646] With the exception of technical quality and creativity, Non-SiA user innovators rated their ideas higher across all categories than SiA user innovators, but only the difference in benefits to others (today) were shown to be significant with $p < 0.05$.[647] The generally lower ratings for newness, benefits to others, and sales potential could be related to the fact that older individuals rate the benefits of new ideas lower, because they have less time to withdraw them (for details see chapter 8.4 below).[648] Ratings for creativity are almost identical across age groups and technical quality is the only characteristic where SiA user innovators score higher.

While the average ratings do not suggest major differences among age groups, the distribution of responses provides a clearer picture. Across all quality categories, with the exception of creativity, the share of at least high agreement (ratings of 4 and 5 on the five-point rating scale) with the stated quality are much higher for Non-SiA user innovators. The multiplier between the groups is between 1.4 and 2.1. The difference

---

[645] Franke & Shah 2003 rely also on user self-ratings of their innovations.
[646] Franke & Shah 2003 and Lüthje et al. 2005 had average ratings (on a seven-point scale) for newness of 3.56 / 3.49 with 14.5 % / 24.1 % high agreement, for technical quality of 2.61 with 12.9 % high agreement and for sales potential 3.44 / 4.32 with 24.2 % / 31.2 % high agreement.
[647] Tested via t-test for equality of means in PASW Statistics 18. See Appendix 9 for detailed results.
[648] See Lévesque & Minniti 2006. This interpretation was also confirmed by expert interview #6.

is most obvious for newness, where more than twice the number of Non-SiA user innovators rate their innovations at least as high (26 % vs. 12 %). SiA user innovators also seem to focus on ideas that are more specialized and tailored to a specific problem. Only 28 % agree with the statement that other users would also benefit from their idea, which is well below the 52 % agreement of Non-SiA user innovators to this statement.

Table 31: User Self-Ratings of Innovation Qualities

Innovation Quality	Mean			Innovations with high or very high agreement (in %)		
	Total	Silver Age group	Non-Silver Age group	Total	Silver Age group	Non-Silver Age group
Newness[a]	2.70	2.52	2.68	23.3	12.0	25.6
Technical Quality[b]	2.42	2.58	2.38	19.6	11.5	21.4
Creativity[c]	3.23	3.15	3.20	34.0	34.6	33.9
Benefit to others (today)[d]	3.42	3.00	3.51	47.6	28.0	51.7
Benefit to others (future)[d]	3.44	3.17	3.50	48.7	33.3	51.7
Sales potential (today)[e]	2.84	2.67	2.89	29.4	20.8	31.1
Sales potential (future)[e]	2.97	2.79	3.03	34.3	25.0	36.1

N = 157. Self-ratings on a five-point rating scales were used.
a) 1 – small improvement / modification of existing product; 5 – totally new product
b) 1 – low-tech solution / known technology; 5 – high-tech solution / new technology
c) 1 – not original at all; 5 – very original
d) 1 – very low; 5 – very high
e) 1 – a few; 5 – many

7.3.2 Impact of Motivational Factors on the Innovation Characteristics

All respondents who indicated that they innovated were asked for their motivation to do so. The relationships between motivational factors and the process qualities, innovation types, and innovation qualities (see chapter 7.3.1 above) are analyzed through correlation analysis. Although correlation analysis itself does not allow researchers to draw conclusions on causal relationships,[649] existing theory provides sufficient evidence that motivation affects "*all aspects of activation and intention*"[650]. Therefore, one can safely assume that changes in motivation influence innovation outcomes, if there exists a significant correlation. The significance of the difference between correlation coefficients of age groups is calculated using the Fisher z-transformation.[651]

---

[649] Cf. Bortz & Schuster 2010, p. 160.
[650] Ryan & Deci 2000, p. 69.
[651] See Bortz & Schuster 2010, pp. 160ff.. Calculation of z-value to evaluate statistical significance of difference as follows: $z = \frac{Z_1 - Z_2}{\sqrt{\frac{1}{n_1-3} + \frac{1}{n_2-3}}}$ with $Z_i = \frac{1}{2}\ln\left(\frac{1+r_i}{1-r_i}\right)$.

Table 32: Correlation Coefficients of Motivators with Process Qualities, Innovation Types, and Innovation Qualities

	EXTR. MOTIVATOR			INTRINSIC MOTIVATORS								
	Earn Money			Personal Usage			Reputation			Fun		
Innovation Age	r^{SIA}	r^{NonSIA}	$p^{\Delta r}$	r^{SIA}	r^{NonSIA}	$p^{\Delta r}$	r^{SIA}	r^{NonSIA}	$p^{\Delta r}$	r^{SIA}	r^{NonSIA}	$p^{\Delta r}$
Development Stage	0.380**	0.111 n.s.	0.197	0.193 n.s.	0.079 n.s.	0.603	0.209 n.s.	0.163	0.834	0.139 n.s.	-0.032 n.s.	0.529
Development Time	0.279 n.s.	0.096 n.s.	0.395	-0.025 n.s.	-0.029 n.s.	0.984	0.257 n.s.	0.267***	0.960	-0.141 n.s.	-0.079 n.s.	0.779
Developm. Frequency	0.257 n.s.	0.224*	0.873	-0.103 n.s.	0.032 n.s.	0.549	0.274 n.s.	0.278***	0.984	0.266***	0.088 n.s.	0.412
Cooperation (Ideation)	-0.175 n.s.	0.038 n.s.	0.337	0.140 n.s.	0.043 n.s.	0.660	-0.045 n.s.	0.071 n.s.	0.603	-0.6·9***	0.028 n.s.	0.001
Cooperation (Realizat.)	0.010 n.s.	0.111 n.s.	0.653	0.000 n.s.	-0.112 n.s.	0.617	0.314 n.s.	-0.037 n.s.	0.105	0.000 n.s.	-0.149 n.s.	0.503
Comfort Improvement	0.189 n.s.	0.039 n.s.	0.497	0.115 n.s.	0.325**	0.322	0.295 n.s.	0.061 n.s.	0.280	0.277**	0.230*	0.826
Cost Reduction	-0.125 n.s.	-0.056 n.s.	0.757	-0.277 n.s.	0.056 n.s.	0.129	-0.359*	0.025 n.s.	0.074	-0.054 n.s.	0.133 n.s.	0.308
New Functionality	0.087 n.s.	0.005 n.s.	0.711	0.366*	-0.173*	0.013	0.080 n.s.	-0.022 n.s.	0.646	0.213 n.s.	-0.117 n.s.	0.136
Time Savings	-0.125 n.s.	0.125 n.s.	0.263	-0.249 n.s.	-0.012 n.s.	0.280	0.080 n.s.	-0.115 n.s.	0.384	0.003 n.s.	0.051 n.s.	0.849
Improved Compatibility	-0.088 n.s.	-0.062 n.s.	0.904	0.316 n.s.	0.150 n.s.	0.430	0.428**	0.031 n.s.	0.057	0.214 n.s.	0.101 n.s.	0.603
Newness	0.085 n.s.	0.253**	0.441	0.132 n.s.	0.042 n.s.	0.689	0.445**	0.047 n.s.	0.054	0.306 n.s.	0.093 n.s.	0.322
Technical Quality	0.319 n.s.	0.236**	0.689	0.097 n.s.	-0.006 n.s.	0.646	0.003 n.s.	0.088 n.s.	0.704	0.312 n.s.	-0.060 n.s.	0.087
Creativity	0.038 n.s.	0.212**	0.430	0.266 n.s.	-0.033 n.s.	0.174	0.213 n.s.	0.153*	0.779	0.501**	0.065 n.s.	0.030
Benefits for Others (today)	-0.061 n.s.	0.074 n.s.	0.549	-0.015 n.s.	0.000 n.s.	0.944	-0.412*	0.025 n.s.	0.038	-0.163 n.s.	0.127 n.s.	0.194
Sales Potential (today)	0.054 n.s.	0.058 n.s.	0.984	-0.103 n.s.	-0.082 n.s.	0.928	-0.350*	0.077 n.s.	0.048	0.061 n.s.	0.051 n.s.	0.968

Correlation coefficients according to Spearman's rho
P-values (2-tailed) calculated with Fisher z-transformation

* $p < 0.10$
** $p < 0.05$
*** $p < 0.01$

Results show that extrinsic factors are almost irrelevant as motivators for users. 97 % of respondents stated that they did not want to earn money with their innovation, and 97 % said they did not receive any kind of financial support. In contrast, agreement on intrinsic factors was very high. 98 % stated that they wanted to use the innovation themselves, and 93 % enjoyed the process of innovating. The ratings regarding reputation as a motivator were mixed, with 22 % agreeing and 44 % disagreeing. Relevant differences between age groups were not observed.

Table 32 above summarizes the correlations of the motivators with the characteristics of the innovation process and its outcomes (see also chapter 7.3.1 above). The extrinsic motivation of receiving financial support for the innovation was eliminated from the analysis due to a lack of variation in responses.[652]

Extrinsic Motivation

Users who wanted to earn money with their innovation developed their innovations further and innovated more frequently. Earning money does not correlate significantly with any of the innovation types, but it correlates with a higher quality of the innovation. Newness, technical quality, and creativity are all rated higher when money is a motivating force. Surprisingly, earning money does not correlate with benefits for others or the sales potential of the innovation. This could either suggest that extrinsically motivated user innovators do not evaluate their ideas differently than intrinsically motivated users, or that most of the other innovations could be

---

[652] Out of the 157 innovators, 151 (96 %) strongly disagreed with that statement and only four indicated high or very high agreement.

successfully commercialized (since they have about the same sales potential) if the user innovator would be motivated to do so. No significant differences between age groups were detected.

Intrinsic Motivation

Intrinsic motivators were found to be generally more important than the extrinsic ones.[653] Developing a product for personal usage was only significantly correlated with innovation type (comfort and new functionality). Interestingly, the relationship between personal usage and new functionality was positively correlated for SiA user innovators and negatively for Non-SiA user innovators. Apparently, older people have specific needs that are not fulfilled by existing products, so they decide to innovate themselves. The intrinsic motivators reputation and fun showed the most significant correlations. For Non-SiA user innovators reputation shows weak positive correlations with process qualities (development stage, time, and frequency). For SiA user innovators, reputation seems to be more important. It correlates positively with the newness of innovations and those that focus on improved compatibility. It is moderately negatively correlated with cost reducing innovations and benefits to others, as well as sales potential. This means that, if reputation is the motivating factor, SiAs focus on new ideas that improve compatibility, but they focus strongly on their own needs and do not build with other users' requirements in mind. Finally, fun seems to foster creativity among older user innovators, showing the highest positive correlation in the analysis. On the other hand, it is strongly negatively correlated with cooperation during the ideation phase for SiA user innovators. Apparently, if enjoyment of tinkering is the key driver for innovative behavior, the elderly want to experience that enjoyment by themselves and do not wish to be disturbed by others.[654] Füller, Jawecki, and Mühlbacher (2007) called innovators who enjoyed the activity of innovating itself *excitement-driven innovators*. They contribute frequently, spend more time on their innovation and reach a higher development stage.[655] With

---

[653] Schuhmacher and Kuester (2012) come to the same result in their recent study on the impact of lead user characteristics on idea quality in service innovations. While extrinsic rewards have an insignificant negative impact on idea quality, intrinsic motivation has a significant and strongly positive impact on the quality of ideas. Cf. Schuhmacher & Kuester 2012, p. 437.

[654] Of course, since causality cannot be determined with correlation analysis, another interpretation is also possible: Elderly who cooperate during the ideation and early development phase of an innovation have less fun in the process.

[655] Cf. Füller et al. 2007, p. 65. In their study among members of an online basketball community, they focused on contributions to innovations on basketball shoes. 80 % of innovators were classified as excitement-driven and 20 % as need-driven innovators.

the exception of development stage, the same observations were made in the study's sample.[656]

7.3.3 Impact of Age on Innovation Characteristics

The impact of age on the different characteristics of the innovation process and its outcomes was to be studied exploratively. To do this, all dimensions of the process quality, the innovation type, and the final innovation qualities were compared between different age groups within the measurements described in chapter 7.1.2.1 above.[657] For each age measurement, groups were divided either according to the separating value of SiA (i.e., ≥ 55 years) or by comparing the upper and lower third of respondents. Where applicable, both grouping procedures were applied. The mean differences among the age groups were compared through the t-test for equality of means.[658]

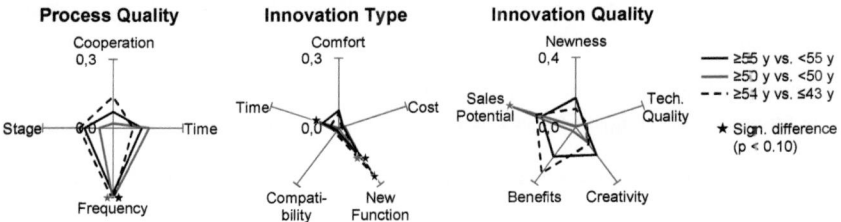

Figure 32: Comparison of Absolute Mean Differences with Different Separators of Chronological Age[659]

The comparison of results across age measurements shows no clear and reliable trend. None of the age measurements is a superior indicator than others, and none of the measurements can indicate all differences in innovation characteristics. The

---

[656] When comparing average scores of process qualities of respondents who strongly agreed with fun being a key motivator with those who disagreed, it shows that they scored higher on development frequency (3.29 vs. 2.73) and development time (2.39 vs. 2.20), and equal on development stage (2.52 vs. 2.55).

[657] Chronological age, Innovation age, Cognitive age and all its sub-dimensions *FEEL* age, *LOOK* age, *DO* age, and *INTEREST* age.

[658] Although the t-test requires normally distributed data, it is robust against the violation of this requirement if group sizes and group variances are about the same size. Cf. Bortz & Schuster 2010, p. 122.

[659] Own illustration. N = 147.
Age groups defined by 1) SiA criterion, 2) age separator (cut off value = 50 years), and 3) upper and lower third. Mean differences tested with t-test for equality of means. Items of process quality and innovation quality were measured with a Likert rating-scale from 1 to 5 (except Cooperation: from 1 to 4). Innovation type was measured through a binary scale (1 - yes, 2 - no).

selection of alternative separating values to create age groups did not yield better results. Figure 32 above shows the results of the application of three different separating values for chronological age. As can be seen, the ranking of the size of the absolute mean difference between groups is not constant. The largest mean difference appears at least once in any of the three group comparisons. Since the gap between the upper and the lower third of chronological ages is largest, it was expected that this comparison would also result in the largest differences. This was only the case in seven out of the 14 possible cases.[660]

Generally, it can be stated that the separation into groups according to age (irrelevant of how age is measured) detects:
1. No differences in process qualities,
2. Only few differences in innovation type (e.g., new functionality more common among younger innovators) and innovation qualities (e.g., lower rating of creativity or benefits for others),
3. Separation according to cognitive age and its sub-dimensions creates greater mean differences but no more significant results.

A summary of significant mean differences for all age measurements is shown in Table 33 below (the detailed figures can be found in Appendix 9).

Table 33: **Summary of Findings Regarding Mean Differences with Respect to Different Age Measurements**

Age Measurement	Compared Age Groups	Significant Mean Differences of Older Age Group Compared to Younger Age Group[a]
Chronological Age	≥55 vs. <55	• Lower share of new functionality innovations (46.5 % vs. 62.5 %) • Lower share of time saving innovations (9.3 % vs. 19.2 %)
Innovation Age	≥55 vs. <55	• Higher development frequency (3.7 vs. 3.1) • Lower rating of benefits for others (3.0 vs. 3.5)
Cognitive Age	≥47 vs. ≤38.3	• Lower share of new functionality innovations (46.9 % vs. 72.5 %) • Lower share of time saving innovations (8.2 % vs. 19.6 %)
FEEL Age	>42 vs. <42	• Lower share of new functionality innovations (44.6 % vs. 69.6 %) • Lower rating of benefits for others (3.3 vs. 3.6)
LOOK Age	>47 vs. <47	• Lower share of new functionality innovations (49.0 % vs. 66.2 %) • Lower share of time saving innovations (9.8 % vs. 20.8 %)
DO Age	>42 vs. <42	• Lower share of cost reducing innovations (3.9 % vs. 12.9 %)
INTEREST Age	>42 vs. <42	• Lower share of new functionality innovations (46.7 % vs. 67.8 %) • Lower rating of creativity (3.0 vs. 3.4)

a) Tested with t-test for equality of means. All mean differences with $p < 0.10$ are reported here.

In a second step, potential differences among older user innovators depending on how old they actually felt were to be analyzed. The idea was that older user innovators who perceive themselves as younger would potentially also innovate more

---

[660] In two of the seven cases, the gap to the second largest mean difference was less than 0.01.

like younger innovators. For this analysis all innovations by innovators who were at least 50 years old at the time of the innovation (innovation age ≥ 50 years) were analyzed. 52 innovations met that requirement. For every innovator, the age difference between overall cognitive age and all sub-dimensions was calculated by subtracting the chronological age. The groups were then divided by the median for each age difference, and mean differences were compared.

Table 34: Summary of Findings Regarding Mean Differences with Respect to Age Differences of Older User Innovators[661]

Age Difference of ... and Chronological Age	Median	Significant Mean Differences of Group that Felt Comparably Older[a]
Cognitive Age	-9.125	- Lower share of comfort improvements (73.1 % vs. 92.3 %) - Higher rating of technical quality (2.8 vs. 2.2)
FEEL Age	-10.0	- Lower rating of development stage (2.0 vs. 2.7) - Lower rating of development frequency (3.3 vs. 3.7)[b] - Lower rating of cooperation (1.0 vs. 1.4) - Lower rating of creativity (3.0 vs. 3.5)
LOOK Age	-5.5	- Lower rating of development frequency (3.3 vs. 3.8)
DO Age	-10.0	- Lower rating of development stage (2.1 vs. 2.7) - Lower share of comfort improvements (72.0 % vs. 92.6 %) - Lower rating of creativity (3.0 vs. 3.5)
INTEREST Age	-8.0	- n/a

a) Tested with t-test for equality of means. All mean differences with p < 0.10 are reported here.
b) Mean difference falls short of required significance with p = 0.103.

In contrast to the previous analysis, the separation of older user innovators according to age difference resulted in many differences between groups along the process qualities (see Table 34 above). Especially the sub-dimensions of *FEEL* age and *DO* age reveal several differences and provide more information than the overall age difference calculated based on cognitive age. Groups split according to *INTEREST* age difference show no significant differences. Groups split according to *LOOK* age show difference only in one characteristic: development frequency.

In summation, the larger the age difference based on the *FEEL* and *DO* age of SiA user innovators, the higher the development stage and the development frequency. They are also more likely to develop comfort innovations, and their innovations will be comparably more creative.

---

[661] For detailed figures, please refer to Appendix 9.

Part C. DISCUSSION & CONCLUSIONS

The third part of this dissertation will summarize the findings and derive recommendations for research and practical applications. Chapter 8 discusses the empirical findings of the studies along the four main research questions. The final chapter, 1, highlights the implications of these findings on the employed research areas of lead user theory and SiMa phenomenon. It also provides recommendations to implement the findings in managerial practice. Finally, the limitations of this study and suggestions for further research into the phenomenon of user innovation in the SiMa are supplied.

8 Discussion

8.1 RQ1: Do User Innovators Exist in the Silver Market Population?

The purpose of this first research question was to ascertain whether user innovators exist across all age groups and whether the share of user innovators would be lower in the SiMa.

The comparison of innovator shares across age groups ranging from below 30 years to over 75 years showed that innovators do exist at all ages. This result confirms initial findings, which were recently introduced by Flowers et al. (2010), Ogawa and Pongtanalert (2011), and Hippel, Jong, and Flowers (2012).

Ultimately, user innovators exist among the SiMa population, and their share is not to be neglected. 39 % of older people reported having innovative ideas (compared to 47 % of younger people) and 23 % reported having built at least a working prototype (compared to 32 %). The innovator shares are lower in the SiMa but are still on very comparable levels. Only the difference for innovators with a working prototype is significant (but on a level of only $p < 0.1$). Reasons for the lower share appear to be brought about by lower extrinsic motivation, limited resources, poorer health and a reduction in cognitive and sensory capabilities. As was shown in chapter 0 above, financial motivators are not important for SiA user innovators. Instead, they innovate more frequently and have to prioritize their projects in order to invest sufficient resources. Their decline in health, cognition, and sensory capabilities is still their largest innovation barrier. This is reflected in the strong correlation between (cognitive) age difference and innovative behavior. The younger an old person perceives himself or herself, the more likely he or she is to develop a promising innovation. The importance of cognitive age as an indicator to segment the SiMa will be discussed in the chapter 8.4 below.

Surprisingly, the innovator shares (independent of whether based on idea or prototype development) peaked at the age group of 50 - 54. This peak is well above the suggested climax of creative output of scientists and research employees. Research on employee behavior and capabilities suggests that this climax should be in the early to late 30s.[662] A possible explanation for this difference could be that the research field requires specific knowledge, which takes time to acquire and master.[663] This is supported by the fact that in the research field of camping and caravanning use experience positively impacts product knowledge. Another possible explanation lies in the fact that the innovations have a high practical relevance. Simonton (1988) and Bergmann, Prescher, and Eisfeldt (2006) found that the typical peak in creative output does not exist if only engineers, who are focusing on immediate product development, are considered. In these cases, creativity seems less important and experience is a major driver.

In summary, the first research question can be answered positively: User innovators exist in the SiMa as they do in younger age groups. The innovator share is slightly lower but differences are not significant. However, SiMa user innovators seem to struggle with the final realization of their innovative ideas, resulting in a significantly lower share of innovators building a working prototype.

8.2 RQ2: Which Determinants of Innovative Behavior Characterize the Silver Market User Innovator? Do these Determinants Differ Compared to Younger User Innovators?

To answer the questions regarding which determinants are typical characteristics for user innovators in the SiMa, the baseline has to be defined first. For this purpose, the most often cited determinants of innovative behavior were analyzed first. After establishing which relationships were relevant and significant, the Non-SiMa and the SiMa populations were compared. Finally, the demographic characteristics of innovators and non-innovators within the SiMa population were compared to explore whether there exist significant differences that would be easy to spot.

---

[662] Cf. Oberg 1960, pp. 251ff.; Simonton 1988, p. 262; Hoisl 2007, p. 21.
[663] Cf. Simonton 1988, p. 252.

8.2.1 General Determinants of Innovative Behavior as a Baseline

The analysis of the general structural model without any separation into age groups confirmed ten of the 15 initial hypotheses. Especially the impact of technical expertise on product knowledge, the lead user components and innovative behavior were confirmed. Product knowledge only shows a positive impact on the ahead of trend dimension of lead userness. The postulated impact on high expected benefits and innovative behavior was not confirmed. Potentially, having enough time, resources, and motivation for a trial-and-error approach is sufficient to make up for a lack of product expertise. A comparable compensation mechanism has been found to exist for technical expertise.[664] After all, although the path from product knowledge to innovative behavior did not show a significant strength, innovators did show a higher average score on product knowledge.[665]

The two lead user components have been found to be moderately correlated, but factor analysis has also shown that both components are clearly independent from each other. This result is in contrast to Morrison, Roberts, and Midgley (2004), who argue that the components are strongly correlated and are of a reflective nature. This supports the statement made by Franke, Hippel, and Schreier (2006) that the two components "*[...] are conceptually independent [...] and they serve different functions in lead-user theory. Although they may be related in some cases and to some degree, this is not necessarily the case.*"[666] Surprisingly, the two lead user components do not have the same importance for innovative behavior. While high expected benefits strongly impact innovative behavior, this is not the case for the ahead of trend component. A relative trend advantage does not automatically lead to innovations. Only when being ahead of trend leads to dissatisfaction and the cognition of deficiencies of existing products, will the impulse to start innovating be sufficient. This confirms the procedural approach of Lüthje (2000), who argues that capabilities, knowledge, and motivations are linked in an overall cognitive process, where, at the beginning, new needs lead to dissatisfaction and, ultimately, to innovative behavior if all other required factors are in place.[667] Schuhmacher and Kuester (2012) found similar evidence analyzing service innovation ideas of lead users, where being ahead of trend had an insignificant, negative impact and dissatisfaction a significant, positive impact on idea quality.[668] According to Franke,

---

[664] Cf. Voss 1985, p. 117; Tietz et al. 2005, p. 336.
[665] Average unweighted construct scores of product knowledge were 4.1 for innovators and 3.7 for non-innovators. Measurement was on a 5-point Likert scale from 1 – very low to 5 – very high.
[666] Franke et al. 2006, p. 303.
[667] Cf. Lüthje 2000, pp. 25f.
[668] Schuhmacher & Kuester 2012, p. 436.

Hippel, and Schreier (2006), innovation likelihood is highly associated with expected benefits, while commercial attractiveness is associated more strongly with being ahead of trend.[669] In the presented example of camping equipment, there exist a high number of innovations, but this does not translate into a high number of commercially successful products.[670] Being ahead of trend correlates more strongly with the innovation qualities than with the high expected benefits. The two lead user components are, in fact, largely independent of each other and relate to different aspects of user innovation.

Use experience had a significant direct impact on product knowledge and innovative behavior, but not on the lead user components. This finding was unexpected. It can be assumed that this finding is rooted in the specific characteristics of the research field of camping and caravanning. As interviews with users have revealed, the main purpose for camping activities is relaxing and recharging personal energy levels. Users typically do not search for increasing excitement or try to push personal boundaries, which has often been the case in previous lead user studies.[671] In contrast to extreme sports in previous lead user studies, camping tourists are looking for relaxation and recreation, and they typically find this in known places and activities.[672] With the exception of camping novices, this is independent from the duration or intensity of the use experience. The resulting impact on the lead user components is, therefore, low. Some camping tourists are trendsetters and can be regarded as lead users in their field – but their amount of use experience does not distinguish them from the rest.

8.2.2 Differences on Determinants of Innovative Behavior between Silver Market and Non-Silver Market User Innovators

The differences of determinants of the innovative behavior are analyzed on four levels. First, the results of the research model of the Silver-Age group are compared against the general structural model to receive an overview of the differences. Second, the results are further refined by comparing the results of the two sub-

---

[669] Cf. Franke et al. 2006, p. 311. Similar in Hippel 2005a, p. 67, but without the clear distinction between the two lead user components.

[670] Only 7.6 % of all innovations are turned into prototypes that are used by more people than just the innovator and only 2.5 % of innovations are commercialized. This finding is also backed up by the evaluation of the company representatives in the first study, which reported a rather low market potential for product ideas suggested by users (see chapter 6.3).

[671] As discussed in chapter 3, most lead user studies in a consumer goods setting focused on extreme sporting communities in which the activities were competitive by definition.

[672] This shows also in the favorite camping activities as collected by Outdoor Foundation 2012, p. 27 among US campers. The vast majority of 76 % mentioned hiking as their favorite activity, followed by cooking and fishing. Activities like climbing, triathlon, rafting, surfing, and snowboarding were not in the top ten and were typically only mentioned by 1 to 10 % of respondents.

samples based on chronological age against each other. Thirdly, results of the separation according to cognitive age are discussed. Finally, the benefits of applying a combination of chronological and cognitive age to separate age groups are discussed.

The comparison of the SEM of the SiMa population with the general SEM reveals three clearly distinguishable differences. Firstly, the SiMa SEM lack the significant impacts of use experience on product knowledge and innovative behavior. The detailed analysis of the characteristics and impact of use experience has shown that the effect is non-linear effect (see chapter 7.2.11 above). The impact of the effect decreases with size. Since older consumers have accumulated much more use experience on average, the impact of changes is much smaller. A similar effect has been identified as *functional fixedness* in the literature.[673] It states that if a user is very familiar with a product, he or she can hardly imagine using it in a different way.[674] Another difference of the SiA group is that product knowledge has a negative impact on high expected benefits. In the general model this relationship is not significant. Slaughter (1993) stated that *"several studies have found that the degree of innovation by users does not depend upon their expertise in the particular field."*[675] Although this statement holds true for the full sample, it does not hold true for the different age groups. The impact of product knowledge on high expected benefits is negative for the SiA group and positive for the Non-SiA group, so that the total effect is not existent. A reevaluation of the studies mentioned by Slaughter (1993) could reveal the same effect. Older people, therefore, seem to expect fewer benefits when they are very knowledgeable about their equipment and market offerings. This could mean that knowledgeable older users know more about alternative products and are better able to identify workarounds to avoid being dissatisfied than their younger counterparts with high product knowledge. This could also be a sign that older consumers suffer more from functional fixedness, which is induced by a decrease in creativity.[676] The more they know about their products, the less they can imagine new uses for them.

The review of the results from the SiA and the Non-SiA group confirms that the research field is still very explorative. Out of the twelve hypotheses derived from the expert interviews, four (H8a, H8b, H9a, H9c) were rejected because there were no actual differences between the age groups. In the case of H13, there was a difference, but it was opposed to the expectations. On the one hand, this shows that

---

[673] Cf. Adamson 1952.
[674] Cf. Alba & Hutchinson 1987, p. 427; Fichter 2005, p. 358.
[675] Slaughter 1993, p. 82.
[676] Cf. Simonton 1988, p. 252.

even experts in the field have a hard time predicting the outcome because there is little sufficient relevant research. On the other hand, this is a sign that the experts were suffering from the Hawthorne effect[677] and were trying to argue for a group difference although "no difference" was a very valid option. The mere question for a potential difference might have motivated them to respond accordingly.

Significant differences exist regarding the impact of use experience on innovative behavior (H8c). As outlined above, this is due to the non-linear effect of use experience and the existence of functional fixedness. As has already been mentioned, the impact of product knowledge on high expected benefits even witnesses a sign change (H9b). While the effect is positive for the Non-SiA group, it is negative among the SiAs. This effect is probably best explained in combination with the stronger impact of being ahead of trend on high expected benefits (the two lead user components) in the SiA group (H11). Among younger users, high product knowledge leads to the recognition of boundaries of the existing market offering and, ultimately, to dissatisfaction with existing products. A relative trend advantage (and the intensive product usage that comes along with it) also leads to this dissatisfaction, but both effects are on comparable levels. The situation is different among older users. For them, a relative trend advantage has a much stronger impact on dissatisfaction with existing products. This is because products must fulfill the high standards of a trendsetter and must additionally cater to the added requirements that come with age. A large product knowledge and good market overview can compensate for a part of this effect, because they can lead to the identification of alternative products that might be better suited. In combination with the additional time and financial resources of older consumers, they can try more options and experiment with potential alternative products before they must admit that none of them are working for them.

The interpretation of the statistical results showed that the lead user components are more strongly correlated among older users. If an older user has a relative trend advantage, he or she will be more dissatisfied with existing products and expect benefits from improvements. The correlation is significantly lower for younger users. The definition of lead users assumes that the two components are independent.[678] According to the findings of this study, this statement is especially true for younger users – among which most of the existing lead user research has been conducted (see Table 2 above). For older users this statement is still true, but to a lesser degree. Their relative trend advantage is much more strongly associated with

---

[677] See Adair 1984.
[678] Cf. Hippel 1986, p. 796; Franke et al. 2006, p. 303.

dissatisfaction, caused by the reasons stated above. The final difference concerns the impact of high expected benefits on innovative behavior (H13). While this impact is very strong among the Non-SiA group, it is significantly lower for SiAs. The potential benefits that a user can gain from using a product innovation motivate older consumer less than younger ones. According to Lévesque and Minnit (2006), older users discount potential benefits from innovations more because they have less time to benefit from them.[679] If the investments in product development are equal for two individuals, the younger one can benefit longer from using the product because he or she has a longer life expectancy.[680] Therefore, the younger one will be more likely to innovate. This theoretical connection also holds true in the empirical data. Although the impact of technical expertise on innovative behavior did not result in any significant difference, its importance still increases among SiAs if interpreted relatively to all other determinants. While in the Non-SiA group, technical expertise had the lowest impact on innovative behavior after expected high benefits and use experience, it is the most important determinant to explain innovative behavior in the SiA group. This can be explained by the stability of technical expertise, which does not decline over time. Technical expertise is crystallized intelligence[681] that is formed during an early formative period in one's teens and early twenties.[682] It is then only marginally affected by decay through aging and becomes therefore a more important and reliable resource during advanced age.

The separation of respondents based on their cognitive age, surprisngly, only resulted in minor group differences. Only the impact of high expected benefits on innovative behavior was lower for the group with high cognitive age. The explanation for this difference is congruent with the one for chronological age: potential profits from innovating are lower for older people than for younger people because they attach a higher discount rate to future profits.[683] No other relationships between respondents with high and low cognitive age yielded significant differences.

Therefore, a segmentation of users based only on cognitive age does not promise relevant results. A 30 year and a 50 year old person who both perceive themselves as 40 years old will most probably still behave differently. Only when cognitive age is considered in relation with chronological age, does it provide interesting insights. The

---

[679] Lévesque & Minniti 2006 aim to explain differences in entrepreneurial behavior but their conclusions are also valid for innovative behavior.
[680] This statement requires the assumption that the time period in which benefits are experienced is not limited, at least not to a time frame shorter than the expected life expectancy of both individuals.
[681] Cf. Horn & Cattell 1967; Sorce 1995, p. 468.
[682] Cf. Becker 2000, pp. 115f.
[683] Cf. Lévesque & Minniti 2006, p. 178.

resulting age difference allows researchers to draw conclusions regarding how an individual might act in comparison to his or her age cohort. Here it is important to state that age difference should only be used to segment members of the same age cohort and should not be used as a universal measure. Again, this can be illustrated with a simple example. Individual A is 30 years old, and individual B 60 years. Both perceive themselves to be each 10 years younger. It was shown that the age difference typically decreases over chronological age (see Figure 23 above). People below 30 typically perceive themselves older. After 30, the age difference gradually decreases until it stabilizes itself around the age of 60 at -8.5 years.[684] In this case, individual B would perceive himself like most individuals in his age cohort and would behave typically. Conversely, individual A acts contrary to expectation by judging himself to be younger. Although both show the same age difference, their values and behavior are probably very different. Additionally, the resulting cognitive age is 20 years for individual A and 50 years for individual B. Again, this shows that their perceived age is very different and not comparable.

The implication that age difference is only a good segmentation variable within a specific age cohort can also be shown in the data. A separation of the full sample (ages 19 to 86) based on age difference has led to only one significant difference and two additional path coefficients, which were significant in only one of the two groups. The separation of the SiMa sample (ages 55 to 86) resulted in three significant differences and two path coefficients, which were significant in only one of the two groups. The differences between groups were generally larger in the SiMa sample.[685]

The prevalent recommendation that cognitive age is a more reliable differentiator than chronological age,[686] therefore, must be specified more precisely. Cognitive age itself is only a good basis for segmentation when one is looking at one age cohort.

---

[684] Comparable results for the age differences have been found in Hubley & Hultsch 1994, p. 416; Cleaver & Muller 2002, p. 238.

[685] The only significant group difference in the full sample was the path coefficient of technical expertise to product knowledge. Additionally the path from use experience to innovative behavior was only significant in the low age difference group while the path from technical expertise to high expected benefits was only significant for the high age difference group. In the Silver Market sample the following path coefficients showed significant differences: product knowledge on ahead of trend, product knowledge on high expected benefits, and technical expertise on product knowledge. Additionally, the path coefficient from technical expertise to being ahead of trend was only significant in the low age difference group while high expected benefits only impacted innovative behavior in the high age difference group. The path coefficient technical expertise on high expected benefits was slightly not significant (p = 0.892), but since the difference to the cut-off value of 0.9 was so marginal it could also be counted as a relevant difference. See Table 44 in Appendix 5 for more details.

[686] To be found for example in Barak & Schiffman 1981; Auken & Barry 1995, p. 108; Szmigin & Carrigan 2001, p. 118; Cleaver & Muller 2002, p. 238.

The absolute size and value of age difference should also be reported for studies in age-related research.

8.2.3 Comparison of Demographic Characteristics of Innovators and Non-Innovators in the Silver Market Population

Demographic factors describe an individual, but it is almost impossible to draw conclusions on concrete behavior using them solely. Consumer research should, therefore, not rely on the interpretation of stable personality traits and demographics. Rather, it should analyze behavior within its specific context.[687] One cannot expect that an innovator in the field of camping equipment will also be innovative in consumer electronics and vice versa. Nevertheless, the characterization of innovators can be valuable in defining selection criteria for future searches for user innovators – either for research purposes or in preparation for lead user workshops according to the lead user method.[688] Furthermore, the explained variance of innovative behavior of the structural model was much lower for the SiA sample.[689] This implies that the established antecedents of innovative behavior do not work as well for older people. There might be additional influencing factors that help to explain the likelihood of innovative behavior in older people.

User innovators are typically characterized as being young, male, well educated, and experienced.[690] The comparison of the demographic characteristics of innovators and non-innovators within the SiMa population revealed that, as a matter of fact, many demographic factors showed no difference between groups. Absolute and disposable income, available time, origin, and marital status did not yield significant differences. The intensity and duration of use experience was comparable. Also, the chronological age (with 62.9 years for innovators and 63.7 years for non-innovators) was almost identical. However, when cognitive age is taken into account, it becomes clear that innovators perceive themselves as much younger than non-innovators (negative age difference of 10.0 years versus 7.2 years).

Expected differences also exist regarding gender, education, and occupation. SiMa user innovators are predominantly male and have a higher education. The cognitive stimulation of day-to-day work seems to positively impact innovative behavior. Innovators are typically still employed (although at an already advanced age) and are less often retired.

---

[687] Cf. Foxall 1995, pp. 280ff.; Lüthje 2004, p. 685; Hoffmann & Soyez 2010, p. 779.
[688] See Herstatt & Hippel 1992; Lüthje & Herstatt 2004.
[689] Silver Age $R^2_{Innovative\ Behavior}$ = 0.209, Non-Silver Age $R^2_{Innovative\ Behavior}$ = 0.367.
[690] Cf. Im et al. 2003, pp. 61f.; Hippel et al. 2011, p. 28.

Summarizing, someone interested in identifying potential innovators in the SiMa population should look for male, well educated, occupationally active individuals with an above average negative age difference to maximize the probability of success. Income, available time, and chronological age by itself are not well suited as selection criteria.

8.2.4 Summary and Response to Research Question

The evaluation of the determinants of innovative behavior shows that there are five distinct differences regarding SiMa user innovators:

1. The impact of use experience on innovative behavior is decreasing with size and is therefore not relevant for SiMa users.
2. Product knowledge impacts the expected benefits negatively for SiMa users, instead of positively, as in the case of younger users, i.e., high product knowledge leads to more product satisfaction for older users and dissatisfaction for younger ones.
3. The lead user components ahead of trend and high expected benefits are more strongly correlated.
4. Expected benefits have significantly less impact on the final innovative behavior.
5. Technical expertise gains relative importance as a determinant of innovative behavior.

Results based on cognitive age do not reveal any more insights. The additional evaluation of demographic factors within the SiMa population showed that significant differences between innovators and non-innovators also exist regarding the following demographics:

- Gender: Innovators are predominantly male.
- Education: Innovators have a higher education.
- Occupation: Innovators are less likely to be retired and are more likely to work full-time.
- Cognitive age: Innovators have a significantly larger age difference.

8.3 RQ3: How Strong - If There Is One - Is the Moderating Influence of Chronological/Cognitive Age on the Determinants of Innovative Behavior?

While the previous research question focused on differences between specific age groups, the third research question investigates whether age measures moderate the impact of the determinants of innovative behavior. To respond to this question, a moderator analysis was conducted. It included all theoretically derived path coefficients influencing innovative behavior and lead user components, but the focus of interpretation was on those that were proven to exist in the general model. Chronological age and cognitive age were each used as the moderating variable. The recently developed two-stage approach created by Henseler and Fassott (2010) was applied.[691]

Chronological and cognitive age both showed a significant positive single effect on use experience, and a negative single effect on technical expertise and being ahead of trend. The positive impact on use experience is quite obvious, because use experience accumulates over time. Therefore, the older an individual, the larger is his or her use experience. The negative impact on technical expertise can be explained by the obsolescence of technical expertise, which is typically accumulated during education and the first years of employment. This effect is also in line with statements from the expert interviews.[692] The negative impact of both age measures on being ahead of trend is relatively weak. Nevertheless, it is not surprising because, as described, older people put more emphasis on security and search for compensating products.[693] They are, therefore, less likely to experiment with new activities and participate in the newest trends. The strength of the path coefficient is always stronger for cognitive age than for chronological age. This strengthens the finding that cognitive age better represents the abilities and attitudes of individuals.[694]

Interpretation of the moderating impact shows that chronological age is not a relevant moderator. Only one path coefficient was influenced by it, and that influence was small in size and significance.[695] Cognitive age moderates more relationships but all

---

[691] See Table 27 for detailed results.
[692] Experts #2 and #4 specifically predicted this outcome.
[693] Cf. Sudbury & Simcock 2009, p. 30; Dychtwald & Flower 1990; Malanowski 2008.
[694] Cf. Auken & Barry 1995, p. 108; Barak & Schiffman 1981, p. 605; Kohlbacher & Chéron 2011, p. 180.
[695] Chronological age moderates the relationship of use experience on innovative behavior ($\gamma_{Chron.\ Age*UE,IB}$ = -.104, p < 0.10) only. Further analysis showed that this relationship is of non-linear nature and that the impact of use experience diminishes. Since chronological age is strongly correlated with use experience the moderator effect can be explained with this non-linear relationship.

are of only low magnitude and have significance levels around 0.10.[696] An interpretation because of the low size and significance does not seem to be meaningful.[697]

Altogether, it can be stated that the moderating influence of chronological age or cognitive age across the full age range is negligible. The significant differences uncovered through the comparison of age groups as described in the previous chapter could not be replicated by the moderator analysis. Two potential explanatory approaches are cogitable. Firstly, the moderator effect could be very gradual. In this case, the accumulated total effect over a long period (e.g., an age range from 20 to 80) could result in noticeable differences between younger and older individuals.[698] Secondly, the differences between the age groups could develop rapidly within a specific age group. That age group could be the 50s, as researchers in the SiMa often argue.[699] Further research with a large, representative sample across all ages should be conducted to test for the latter option.

8.4 RQ4: Do User Innovations by Silver Market User Innovators Differ from "Regular" User Innovations, and if so, How?

While the first three research questions dealt with whether user innovators existed in the SiMa and how they were different, the final research questions tries to shed light on whether there exist differences in the resulting innovations. Insights regarding the following four propositions were sought after:

P1: The innovation process will differ between Silver Market user innovators and younger user innovator, especially with regard to development stage, frequency, and cooperation during development.

P2: Silver Market user innovators will focus on different innovation types, e.g., more on comfort and compatibility and less on time and cost reduction.

---

[696] Cognitive age moderates the relationships of use experience on high expected benefits ($\gamma_{Cogni.\ Age*UE,HEB}$ = -.086, p < 0.10) and innovative behavior ($\gamma_{Cogni.\ Age*UE,IB}$ = -.088, p < 0.10), of product knowledge on high expected benefits ($\gamma_{Cogni.\ Age*PK,HEB}$ = -.069, p = 0.137) and innovative behavior ($\gamma_{Cogni.\ Age*PK,IB}$ = -.071, p = 0.119), and of technical expertise on innovative behavior ($\gamma_{Cogni.\ Age*TE,IB}$ = -.065, p = 0.153).

[697] The subdimensions were also tested for their moderating influence but results were in line with the combined cognitive age.

[698] The standard deviation of chronological age is 11.6 years. Average chronological age is 63.4 years for the SiA group and 43.4 years for the Non-SiA group respectively. The difference is already almost two standard deviations.

[699] Cf. Szmigin & Carrigan 2001, p. 115; Auken et al. 2006, p. 440; Gassmann & Reepmeyer 2006; Kohlbacher & Herstatt 2008a, p. xi; Fisk et al. 2009, p. 8.

P3: Innovations by Silver Market user innovators will exhibit different qualities than those by younger user innovators. They are likely to score lower on newness, technical quality, and creativity but higher on benefits to others and sales potential.

P4: Among Silver Market user innovators, cognitively younger user innovators will exhibit differences related to process quality, innovation type, and innovation quality compared to cognitively older user innovators.

The comparison of process qualities, innovation types, and innovation qualities between innovations from SiMa and Non-SiMa users showed that there are relevant differences.

All innovators seem to be serial innovators (98 % innovated more than once), but older ones innovate more often than younger ones. The share of older innovators who stated that they adapt or improve (almost) all their equipment is 30 %, which is five times larger than the share of younger innovators (6 %). This can be explained through the higher amount of disposable time of older users[700] and is a sign for the higher dissatisfaction of older users with general products that do not cater to their needs.[701] Large differences also exist regarding the development stage of the innovations. Older innovators generally reach a lower stage. Only 41 % of older user innovators built a working prototype, compared to 58 % of younger user innovators. According to Lettl, Herstatt, and Gemünden (2006), imagination capabilities, domain-specific expertise, tolerance of ambiguity, and technological expertise are required to transform an idea into a product.[702] Since age-related cognitive decline reduces the mental abilities, imagination, and tolerance of ambiguity are lower among elderly. Therefore, they have a disadvantage in two important characteristics to successfully develop new products. Additionally Braun and Herstatt (2007) identify legal, market, technological, social, and personal barriers to user innovation, but it remains unknown which barriers are especially important for the individual steps of the innovation process. Legal and market barriers should be irrelevant for user innovators developing for themselves. A lack of financial resources is presumably not a concern for the elderly, as they have accumulated savings.[703] Time constraints might be a valid barrier, since older user innovators require more time to reach a specific development stage than younger innovators. On the other hand, older users

---

[700] SiAs stated to have 7.5 hours per day of disposable time, compared to only 4.4 hours per day for Non-SiAs.
[701] Cf. Schmidt-Ruhland & Knigge 2008, pp. 103ff.
[702] Cf. Lettl et al. 2006, p. 39.
[703] See chapter 2.2.1.

have more time available to overcome time constraints. Since older users are less experienced with the internet,[704] they might be more affected by technological barriers. If they encounter a technological problem, they have less access to potential solutions or expert knowledge and must rely on their own creativity.

The first proposition can, therefore, be confirmed. SiMa user innovators differ regarding development stage and frequency. Differences in cooperative behavior could not be affirmed.

The comparison of innovation types showed that the ranking of potential types is identical. Nevertheless, there exist minor differences regarding the importance of some types. Innovations focused on comfort and compatibility were more important for older users, which is in line with the second proposition. Among the innovation types that were less important, the relative differences between the age groups were very large. Time saving innovations, for example, were less important for older user innovators (11 % compared to 18 %). As described above, older users typically have more time available, so time saving is not as important to them. In contrast, cost reducing innovations were mentioned almost twice as much among older user innovators (11 % compared to 7 %). This latter difference was not predicted. Apparently older users are more cost-conscious than expected. Although their savings are above average, their income is lower. The relative high importance of cost reduction might also be an indicator for the existence of poverty among the elderly.[705]

The second proposition is, therefore, confirmed for the most part. Only the higher importance of cost reductions for older user innovators was not expected.

The comparison of the innovation qualities also reveals differences between the age groups. Older user innovators generally assess their innovations lower. Ratings for newness, benefits for others (today and in the future), and sales potential (today and

---

[704] While over 90 % of the population between 14 and 49 years use the internet, only 24 % of those of 65 years and older do. Cf. BITKOM 2011, p. 9.

[705] A comparison of pension system across OECD countries by Disney & Johnson 2001 attested the German pension system a good protection from poverty with very little income equality. Only small groups were threatened by poverty, e.g., single elderly women, but even these were generally less common than poverty in the general population. (about 4.2 % of single elderly women were considered to be below the poverty line compared to 7 % in the general population) (cf. Disney & Johnson 2001, pp. 186ff.). Due to social reforms in 1999 and the demographic shift, this portion has increased tremendously. The latest comparison of elderly's living conditions in the European Union showed that the share of individuals above 65 years who run the risk of poverty is currently at 15 % in Germany (cf. Haustein & Mischke 2011, p. 66). The reader should note that the sharp increase is partly due to different approaches how to calculate poverty risk. While Disney & Johnson 2001 used eligibility for social assistance (which was then the official definition for the poverty line in Germany), the latest figures are based on the median income within the population (poverty line is defined as 60 % of the median income, including social aid).

in the future) all have lower scores. Lévesque and Minniti (2006) showed conceptually that older individuals are less likely to become entrepreneurs mainly due to the fact that they have less time to withdraw profits. This individual assessment is potentially generalized by older innovators (according to the confirmatory bias, people tend to interpret information in a way to support their own preconceptions), so that they assess the overall sales potential as low and not worthwhile.[706]

Surprisingly, there are no differences regarding the creativity of the innovations. Although several empirical studies have shown that creative output typically peaks in one's 30s,[707] the subjective evaluations did not reflect this. One reason might be that even the average age of the younger group is at 46.7 years, so that some of the decline has already occurred. Additionally, innovating in the hands-on field of camping equipment has more in common with the work of an engineer than with the abstract tasks of an R&D employee. For engineers, the correlation between age and creative output is only marginal.[708] The comparison of evaluations of technical quality showed no real difference in average ratings, but the scores among younger user innovators were more evenly distributed.[709]

The alignment of innovators' self-evaluations with the assessment of the company representatives shows that newness and creativity are judged equally. Deviations exist for technical quality and sales potential. Both qualities are evaluated much lower by the company representatives. This is in line with the fields of expertise of users and manufacturers. As stated by Hippel (1976b), users typically dominate the early phases of product development, including identification of unmet requirements and building of first prototypes. Manufacturers typically only step in when a product concept is promising and then focus on improving reliability and preparing for commercialization.[710] In areas where manufacturer's expertise is high, they rated the innovations more negatively than in areas where users typically have more expertise.

Only parts of the third proposition could be confirmed. While there are differences in the innovation qualities between the age groups, the direction of the differences was not always as expected.

---

[706] Cf. Plous 1993, p. 233.
[707] Cf. Oberg 1960, p. 253; Adenauer 2002, p. 42.
[708] Cf. Bergmann et al. 2006, p. 25; Oberg 1960, pp. 253ff.
[709] Kurtosis$_{Silver\ Age}$ = -0.588; Kurtosis$_{Non\text{-}Silver\ Age}$ = -1.013. A negative kurtosis indicates that the distribution is flatter than a normal distribution. The larger the difference the flatter the distribution. Cf. Hair et al. 2008, p. 35.
[710] Cf. Hippel 1976b, pp. 220f., 2005a, p. 72.

In addition to the differences between younger and older user innovation, it is worthwhile to mention that the definition of a clear-cut age limit is almost impossible. Depending on the age measurement and the threshold value, results of the analysis can differ widely. In search for a superior age limit, the sample was split based on several cut-off values (50 years, 55 years, median age, and upper vs. lower third) for chronological age and cognitive age, including all its sub-dimensions. The resulting differences were very inconsistent and not reliable. A definition of the SiMa segment based solely on age must therefore be considered as problematic and only suitable for a practice-oriented approach. There is no natural age limit that causes a change in behavior and explains the differences. Other underlying factors (like the changes of cognitive capabilities, physical fitness, and social status) are the root cause for the observed effects.[711] Age, although strongly correlated, is only an indicator and can therefore only act as a proxy.

Chronological and cognitive age can still add to an understanding of user innovators. In combination, both age measures prove to be a good separator when a delimited age group is examined. The separation of the SiMa segment based on the relative age difference has resulted in reliable differences. Besides the age difference based on total cognitive age, the differences based on *FEEL* and *DO* age were especially adequate. Innovators who felt especially young scored better on the process qualities. They reached a further development stage, developed more frequently and cooperated more. Additionally, they rated their creativity higher. The *FEEL* age dimension reflects the emotional age dimension.[712] According to Hubley and Hultsch, feeling younger is associated with extraversion, openness to experience, and an internal locus of control.[713] Since, according to lead user theory, the locus of control "*[...] is an important antecedent of consumers' creativity in problem-solving contexts*"[714], feeling especially young indicates the existence of important innovator capabilities. The *DO* age dimension relates to the societal age dimension.[715] User innovation research has shown that the largest positive external impact typically comes from communities.[716] The innovators in this study, who think that they act especially young, reach a further development stage, develop more comfort improvements, and rate their creativity higher. This is in line with the existing

---

[711] Cf. Moschis 1992a, p. 18; Kohlbacher & Chéron 2011, p. 179.
[712] Cf. Barak 2009, p. 3.
[713] Cf. Hubley & Hultsch 1994, p. 434, 1996, p. 495.
[714] Schreier & Prügl 2008, p. 337.
[715] Cf. Barak 2009, p. 3.
[716] Cf. Franke et al. 2006, pp. 312f.; Baldwin et al. 2006, p. 1307; Ogawa & Pongtanalert 2013, pp. 42ff.

research on community innovators, but further research would be required to confirm equivalence.

What was previously shown in the case of determinants of innovative behavior also holds for the analysis of innovation characteristics: The definition of groups according to relative age difference provides better results than those according to absolute chronological or cognitive age. The fourth proposition can, therefore, be confirmed. Age differences should not only be calculated regarding total cognitive age but also regarding its sub-dimensions; otherwise researchers and practitioners might miss important insights.

9 Contribution and Implications

This chapter relates the findings of this study to existing research and highlights the study's contributions to and implications upon academic research. This is followed by recommendations for managers and practitioners to implement the findings into concrete actions. Even though the research was carried out with the due thoroughness and meticulousness, there are some limitations to the study, which must be highlighted. Lastly, ideas for further research are suggested.

9.1 Contributions to Academic Research

This study contributes to the existing knowledge of the lead user theory as well as to the SiMa theory. Furthermore, the study can contribute to the methodology through implications for the measurement of age through cognitive age and the resulting age difference.

9.1.1 Implications for Innovation Management

This study contributes to the field of innovation management by providing the first quantitative study on the relationship of age and innovation management that compares the innovative behavior of young and old age groups within one product category. The study was also not limited by an age cap of 65 years, which applies to almost all studies in organizational research and human resource management, as that is the typical age of entry into retirement. The study therefore provides some of the first academic insights into the characteristics of age-related changes to innovative behavior that are not job-related.

It was shown that user innovations do not only exist in specific product categories that attract only few individuals but also exist in a mass market.[717] Therefore, they can be involved in the development process for any product and contribute positively to the overall economy and social welfare.

Concerning the lead user concept, the study could confirm that the two lead user components, being ahead of trend and expected high benefits, are independent characteristics that should be measured and interpreted separately in innovation management research.[718] The significantly different correlations between the two components in the analyzed age groups indicate that the degree of independence is not related to stable personality traits. Instead, changes in needs and preferences,

---

[717] See also the articles of Hippel et al. 2011 and Hippel et al. 2012 on consumer innovations.
[718] Cf. Franke et al. 2006, p. 311.

which change with age, as has been previously described, also seem to affect the independence of the lead user components.

The absolute and relative importance of the drivers for innovative behavior differed between the analyzed age groups. Unlike younger users, SiAs' innovative behavior is not impacted by use experience at all, less impacted by expected benefits, and technical expertise gains relative importance towards all determinants. Since demographic factors were controlled, the differences were caused by personal values, needs, and preferences. The study could show that age and the age difference based on cognitive age act as separators for the impact of innovation drivers. Therefore, user innovators differ and are not all alike. Although the key drivers of innovative behavior have been identified in previous studies, not all of them apply to every innovator.

Additionally, this research has shown that the drivers of innovative behavior are not necessarily linear effects, as is typically assumed. In the case of use experience, it was shown that, while at first use experience is required to successfully become innovatively active, its importance decreases over time. Too much use experience seems to lead to functional fixedness, so that the individual might have trouble coming up with new and creative solutions to problems or by identifying problems in the first place due to adaption. In this case "more" does not necessarily mean "better". The potential existence of non-linear effects should be considered when conducting research on the characteristics of user innovators, especially when they might possess a high degree of experience or knowledge.

9.1.2 Implications for Silver Market Theory

Although there are many studies on the demographic change and the psychological, cognitive, and biological effects of aging on humans, the research on the SiMa from the perspective of product development and innovation management is still very limited. The SiMa is often just regarded as another market segment and recommendations are driven by qualitative findings from separate case studies.[719] This study is the first empirical study that links age with user innovation and incorporates all age groups to draw comparisons. It responds to the call for further research by Bogers, Afuah, and Bastian (2010) who demanded to "*[...] explore how the cognitive limitations [...] of economic actors affect their decision-making capabilities in the process of innovation.*"[720]

---

[719] See for example Arnold & Krancioch 2011; Pettigrew 2008; Enomoto 2011; Schmidt-Ruhland & Knigge 2008; Pfeiffer 2011; Bullinger et al. 2011.
[720] Bogers et al. 2010, p. 866.

It was shown that older users are almost as innovative as younger ones in the area of low-tech consumer goods. The resulting innovations differ slightly, especially with regards to their intended purpose. SiA user innovators put their emphasis on innovations, which solve age-specific problems, e.g., more comfort and better compatibility with older equipment, but do not limit themselves to those.

The collection of data has shown that, at least currently, SiAs are less online-affine. They apparently utilize the resources of online communities less than Non-SiAs. This effect will most probably disappear over time, as members of the younger age cohorts, who are already familiar with online communities will grow older - unless their cognitive capabilities prevent them from interacting with these communities. Nevertheless, researchers, who are currently interested in conducting age-related consumer research, cannot only rely on the convenience of online surveys because they will most probably miss large parts of their targeted sample. Instead, the studies have to be conducted at the point of sales and / or consumption in order to collect representative data, as was also highlighted in one of the expert interviews:

"What we do not do is a laboratory situation - like product tests or user test - but we start on-site with a 1-on-1 situation. We look in their homes at the place of usage: how are products handled? How is it used and in which setting? And then we have the user, too. And then, we draw conclusions on the one hand from the discussion with the user, and on the other hand from the observation."[721]

The separation of the SiMa, starting at the age of 55 years, proved to be meaningful, although the definition of the SiMa based solely on a chronological age threshold remains to be difficult. Chronological age is an easy way to define the SiMa, but it's rather imprecise. The additional information contained in cognitive age makes this measure more meaningful, especially when interpreted in comparison with chronological age.

9.1.3 Implications for Measuring Age in Innovations Research

Based on this study, some suggestions and recommendations for measuring age and comparing age groups in organizational and innovation research can be derived.

First of all, if the relationship between age and innovation is in the research focus, the true age at innovation must be considered. Even the last innovation can have occurred several years in the past, so that the analysis would be biased if not corrected.

---

[721] Expert interview #3.

Secondly, when using PLS-SEM, the impact of age should be measured through an MGA and not through moderator analysis alone. Incremental changes can sum up over long periods, so that the comparison of two or more age groups will more likely exhibit existing differences. The challenge for the researcher lies in the definition of valid and comprehensible values for the group separators.

Thirdly, cognitive age itself does not reveal interesting insights on a person. Splitting groups according to cognitive age does not create more interesting results than the split based on chronological age. Nevertheless, cognitive age is a very good measure when analyzed in contrast to the actual chronological age of the respondents. The absolute size and the sign of the resulting age difference in combination are a very good measure to separate members of an age cohort and to classify them.

9.2 Recommendations for Managerial Practice

The findings and contributions of this research provide useful insights for the management of consumer goods companies, and potentially, beyond. As mentioned in the introduction, the question of whether user innovations can improve the development and marketing of products for the SiMa is especially relevant for practitioners in the areas of innovation and marketing management. Additionally, recommendations for overall strategic management can be deduced by providing insights on the existence and characteristics of SiA user innovators.

Based on the results of the study, the following recommendations for managerial practice are formulated:

Involve older user (innovators) in the product development process – but differently!

The analysis has shown that SiA user innovators exist and that they have specific ideas on how to improve their products. Therefore, they can be integrated in ideation and product development processes. The survey among camping vehicle and equipment manufacturers has uncovered that companies are still hesitant to involve users despite the proven benefits.[722] Different innovation management methods have been introduced to integrate users in the innovation process. Beyond simple customer surveys and prototype testing, there exist the lead user method[723], innovation toolkits[724], and virtual user communities[725], amongst others. All these

---

[722] Cf. Hippel et al. 1999, p. 56; Herstatt & Hippel 1992, pp. 219–220; Franke et al. 2006.
[723] See Hippel 1988, pp. 102ff.
[724] See Franke & Hippel 2003.

methods can be used to integrate older users as well. As was shown, SiA users have ideas for new products and product improvements and can articulate their specific needs. But they need support in developing concrete plans and prototypes because the realization of an idea seems to be more problematic to them. Only 48 % of SiA user innovators transform their idea into a prototype, compared to 67 % of younger ones. Product suggestions coming from older users should, therefore, be more carefully evaluated so as not to discredit them solely due to a lack in technical quality or a thorough description.

The success of many of the innovation management methods depends on the successful identification of suitable participants, especially for methods in which the number of participants is limited and the initiator cannot rely upon self-selection alone, e.g., for the lead user method. Based on the findings, companies should look for predominantly male, well-educated individuals with a below average age difference (especially based on feel and do age) and sufficient (but not excessive) experience in the field. Of course, companies should focus on, but not limit themselves to, participants fulfilling these criteria, because otherwise they run the risk of developing for a market niche only.

Silver Market user innovations indicate what is needed by older users.

In order to develop successful products, companies need to know about the specific needs and requirements of their customers. Many efforts have already been made to try to describe and categorize the specific characteristics of the SiMa from a psychological, sociological, and business perspective.[726] This study reveals that there exist differences in the reasons regarding problems SiA user innovations try to solve for. SiA user innovators focused more often on solutions that would increase comfort, improve compatibility, and reduce costs. Professional solutions in these areas are therefore also the most promising. Comfort improvements relate directly to a lower physical constitution, including lower strength, stamina, and stretchability. Manufacturers should incorporate these limitations of older consumers by considering the principles of universal design.[727] The need for a better compatibility arises from the equipment which SiA users already own. In order to fix, improve, or expand them, they demand products that are compatible. Manufacturers should consider this requirement already during the design phase. The use of industry standards and modular design, including a comprehensive documentation, should help to offer products that are compatible with predecessor products – especially in a

---

[725] See Herstatt & Sander 2004.
[726] See for example Bengtson et al. 2009a; Arnold & Krancioch 2011; Usui 2011; Reinmöller 2008; Tempest et al. 2008.
[727] Cf. NC State University 1997; Gassmann & Reepmeyer 2008.

low-tech industry. Finally, the need for cost reductions allows for two conclusions. Firstly, this indicates that SiA consumers pay attention to their budget despite the fact that they typically have accumulated considerable wealth. Their income is lower than during working life, and so they also try to reduce expenses. Secondly, poverty among the elderly plays an increasingly important role.[728] In both these cases, SiA consumers demand products that help them save money in the future. Products that fulfill this requirement will also find a market among younger age groups.

Apply cognitive age and the age difference to segment the Silver Market.

Innovation and marketing experts have reached the conclusion that "*[...] the silver market is by no means a homogenous market segment*"[729]. Nevertheless, the best methods for the segmentation of the SiMa remain under discussion because measures must be easily collectable, reliable, and effective in defining the segments. It was shown that cognitive age and the resulting age difference can be used to segment the market. The data gathering in surveys is simple and not cognitively exhausting. The age difference provides especially interesting insights if the underlying age cohort is narrowly defined which makes it best suited for analyzing the SiMa or even smaller age segments. The sub-dimensions of *feel* and *do* age (relating to emotional and societal aspects of the individual) seem to be the ones that indicate difference best and should be analyzed besides the overall age difference based on the average cognitive age.

Communication channels must be tailored to reality of older users.

Although newest data on online and social media usage indicates that the number of silver surfers is increasing,[730] their online activity is still lower than that of younger age groups. This was evident in the participation in the online communities which was dominated by participants below the age of 50 years. Since SiA users are less online-affine, it is more difficult to approach them via online surveys or in online communities. Companies that want to cooperate with older users must, therefore, approach them where older users actually use their products.[731] Companies can also support older users in voicing their ideas and opinions online by providing specific forums and contests,[732] but they should not only rely on these tools because they are likely to miss large parts of the market.

---

[728] See Haustein & Mischke 2011, pp. 53ff.; Disney & Johnson 2001.
[729] Kohlbacher & Herstatt 2011a, p. xx.
[730] Cf. Saul 15/01/2014.
[731] See Schmidt-Ruhland & Knigge 2008 for a description of their design approach in the *sentha* project (Everyday Technology for Senior Households).
[732] Cf. Bullinger et al. 2011; Leyhausen & Vossen 2011.

The difference in the online and social media usage will certainly become equal over time. Until then, those who are conducting surveys in online communities must pay attention to the fact that silver surfers are not representative for the complete SiMa.

Evaluate whether your own employees have innovative ideas.

Schweisfurth (2013) has shown that lead users also exist inside a company. These *embedded lead users* (ELUs) do not have to be related to the marketing or development department but can still provide valuable input by applying their knowledge, experience, and ideas. Since SiA users have shown to still be innovative, older employees could be a profitable source of innovation as well. By becoming older and by experiencing the specific needs and requirements that come along with age at first hand, they might become an ELU for age-based products. It is also easier and economical to involve own employees.

Companies wanting to assess whether there are ELUs for the SiMa must consider that, due to the typical age of retirement of 60 to 67 years, this expertise is always on the verge of leaving the company. Potential ELUs should therefore be addressed before they retire or the company should try to keep in contact so that they can involve them at a later stage.

Generally, product managers should carefully plan their approach when developing products targeted at the SiMa. The integration of user innovators helps to identify the specific requirements of this market and to efficiently develop successful products.

9.3 Limitations and Suggestions for Further Research

Although this research study was based upon the latest theoretical and empirical findings and the research model was carefully designed, one still has to observe the limitations of this study. The limitations originate mainly from the research methods and the sample characteristics. Suggestions for further research to counteract some of the limitations, as well as some suggestions, which are based on the implications detailed above will be provided.

First of all, since this was the first study that focused on the relationship of age and user innovation, further comparable studies are required to confirm the findings. The sample of the study was collected among camping tourists. Therefore, the sample focused on a low-tech industry, and the results should be interpreted within this context. Although the camping equipment industry is representative for many consumer goods product categories, there are other industries with very different characteristics. In fast-changing industry contexts that require more technical knowledge or specific expertise, e.g., consumer electronics, application development, etc., age might have a different impact on the overall innovative behavior and the required innovator characteristics. Further studies in different industry contexts, e.g., the high-tech industry or fast-moving consumer goods, are required to analyze which findings regarding the impact of age on innovative behavior remain stable and which are industry-specific.

Additionally, the sample is not perfectly representative of the whole population. There are slight differences regarding the age distribution of Germany (no respondents were below 19 years or above 86 years of age), and women are generally underrepresented. While the difference in the age distribution can be neglected, the potential peculiarities of female user innovators are worth analysis. Many other studies have found that innovators are predominantly male, but most of these studies also focus on male-dominated product categories.[733] Claßen (2012) has shown in her study on the technology acceptance of elderly users that the largest differences between age cohorts exist among women.[734] A detailed analysis on the impact of age on the innovative behavior of women, therefore, seems promising.

Research on age always faces the issue that the findings can result from age, period, or cohort effects. The period effect does not exist in the study, because the data was collected at one point in time only. The age and cohort effects, in contrast, cannot be separated. The age effect depends on the time that has passed since birth. The cohort effect depends on the birth date itself.[735] In order to separate the two effects, a long-term study, in which members of the same age cohort are surveyed repeatedly, would have to be conducted.

None of the SiA respondents in the sample stated that they developed their innovation completely on their own. They especially received help from others during the ideation phase and over 30 % also received it during the realization. SiA user innovators also stated that they were mainly motivated by the reputation they could gain from a successful innovation. Due to the scope of this thesis, the influence of communities on innovative behavior of SiA user innovators could not be elaborated further. Since many findings indicate the positive impact of communities on user innovations,[736] their specific impact on SiA user innovators should be researched to determine how (online) communities must be designed, so that SiA users can collaborate best.

---

[733] Cf. Tietz et al. 2005, p. 327; Franke et al. 2006, p. 305; Franke & Shah 2003, p. 162.
[734] Cf. Claßen 2012, pp. 245ff.
[735] Cf. Holford 1983, p. 311.
[736] See Herstatt & Sander 2004; Franke & Shah 2003; Marchi et al. 2011; Janzik 2012.

It was shown that age difference provides valuable insights into the innovative behavior within one age cohort. Since this measure approximates underlying emotional, biological, societal, and intellectual aspects of individuals,[737] it can most probably also support the understanding of other relationships with regard to the SiMa and innovation management. Researchers should apply the age difference to better understand technology acceptance of SiA users or the diffusion of products within the SiMa.

Generally, further research in innovation management is required to explain how innovative behavior and expertise in other domains are related. Concepts like ELUs already highlight the potential benefit for companies of employees and users who combine a high and specialized product knowledge with functional knowledge, e.g., in marketing.[738] It would be interesting to determine whether the domain-specific knowledge also influences the innovation type. For example, in the case of sports, would an engineer be more likely to develop a product innovation? Would a sports scientist be more likely to develop a new (training) technique?

In order to provide manufacturers and policy makers with guidelines to promote user innovations, one must understand which innovation barriers prevent successful user innovations and how to overcome them. While several innovation barriers are already identified, it is currently unknown how they map along the innovation process. The share of innovators who build a working prototype is already low, and only a small fraction of innovators eventually sees their idea being commercialized. To provide user innovators with the best support at each process step, a mapping of barriers to innovation along the process is required.

---

[737] Cf. Barak 2009, p. 3.
[738] Cf. Schweisfurth 2013, pp. 168f.

References

Abraham, W. Todd; Russell, Daniel W. (2004): Missing Data. A Review of Current Methods and Applications in Epidemiological Research. In *Current Opinion in Psychiatry* 17 (4), pp. 315–321.

Adair, John G. (1984): The Hawthorne Effect. A Reconsideration of the Methodological Artifact. In *Journal of Applied Psychology* 69 (2), pp. 334–345.

Adamson, Robert E. (1952): Functional Fixedness as Related to Problem Solving. A Repetition of Three Experiments. In *Journal of Experimental Psychology* 44 (4), pp. 288–291.

Adenauer, Sibylle (2002): Die Älteren und ihre Stärken - Unternehmen handeln. In *Angewandte Arbeitswissenschaften* (174), pp. 36–52.

Agresti, Alan; Finlay, Barbara (1997): Statistical Methods for the Social Sciences. 3rd ed. Upper Saddle River, NJ: Prentice Hall.

Alba, Joseph W.; Hutchinson, J. Wesley (1987): Dimensions of Consumer Expertise. In *Journal of Consumer Research* 13 (4), pp. 411–454.

Albers, Sönke; Hildebrandt, Lutz (2006): Methodische Probleme bei der Erfolgsfaktorenforschung. Messfehler, formative versus reflektive Indikatoren und die Wahl des Strukturgleichungs-Modells. In *Zeitschrift für betriebswirtschaftliche Forschung* 58 (1), pp. 2–33.

Alesina, Alberto; Fuchs-Schündeln, Nicola (2007): Good-Bye Lenin (or Not?). The Effect of Communism on People's Preferences. In *American Economic Review* 97 (4), pp. 1507–1528.

Allen, Robert C. (1983): Collective Invention. In *Journal of Economic Behavior & Organization* 4 (1), pp. 1–24.

Ando, Albert; Modigliani, Franco (1963): The 'Life Cycle' Hypothesis of Saving. Aggregate Implications and Tests. In *American Economic Review* 53 (1), p. 55.

Anstey, Kaarin J.; Lord, Stephen R.; Smith, Glen A. (1996): Measuring Human Functional Age. A Review of Empirical Findings. In *Experimental Aging Research* 22 (3), pp. 245–266.

Arnold, Gunnar; Krancioch, Stephanie (2011): Current Strategies in the Retail Industry for Best-Agers. In Florian Kohlbacher, Cornelius Herstatt (Eds.): The Silver Market Phenomenon. Marketing and Innovation in the Aging Society. 2nd ed. Berlin, Heidelberg: Springer, pp. 149–159.

Astor, Michael (2000): Innovationsfähigkeit, Wissenskulturen und Personalstrategien. In Annegret Köchling, Michael Astor, Klaus-Dieter Fröhner, Ernst Andreas Hartmann, Tanja Hitzblech, Jasper Gerda, Josef Reindl (Eds.): Innovation und Leistung mit älterwerdenden Belegschaften. München: Rainer Hampp Verlag, pp. 317–360.

Auken, Stuart van; Barry, Thomas E. (1995): An Assessment of the Trait Validity of Cognitive Age Measures. In *Journal of Consumer Psychology* 4 (2), pp. 107–132.

Auken, Stuart van; Barry, Thomas E. (2009): Assessing the Nomological Validity of a Cognitive Age Segmentation of Japanese Seniors. In *Asia Pacific Journal of Marketing and Logistics* 21 (3), pp. 315–328.

Auken, Stuart van; Barry, Thomas E.; Bagozzi, Richard P. (2006): A Cross-Country Construct Validation of Cognitive Age. In *Journal of the Academy of Marketing Science* 34 (3), pp. 439–455.

Backes, Gertrud; Clemens, Wolfgang (2008): Lebensphase Alter. Eine Einführung in die sozialwissenschaftliche Alternsforschung. 3rd ed. Weinheim: Juventa-Verlag.

Backhaus, Klaus; Erichson, Bernd; Plinke, Wulff; Weiber, Rolf (2011): Multivariate Analysemethoden. Eine anwendungsorientierte Einführung. 13th ed. Berlin, Heidelberg: Springer.

Backhaus, Peter (2008): Coming to Terms with Age. Some Linguistic Consequences of Population Ageing. In Florian Coulmas, Harald Conrad, Annette Schad-Seifert, Gabriele Vogt (Eds.): The Demographic Challenge. A Handbook about Japan. Leiden: Koninklijke Brill NV, pp. 455–471.

Bailey, Kenneth D. (1994): Methods of Social Research. 4th ed. New York, NY: Free Press.

Baker, George T., III.; Sprott, Richard L. (1988): Biomarkers of Aging. In *Experimental Gerontology, Special Issue Biomarkers of Aging* 23 (4–5), pp. 223–239.

Baldwin, Carliss; Hienerth, Christoph; Hippel, Eric von (2006): How User Innovations Become Commercial Products. A Theoretical Investigation and Case Study. In *Research Policy* 35 (9), pp. 1291–1313.

Barak, Benny (2009): Age Identity. A Cross-Cultural Global Approach. In *International Journal of Behavioral Development* 33 (1), pp. 2–11.

Barak, Benny; Gould, Steven (1985): Alternative Age Measures. A Research Agenda. In *Advances in Consumer Research* 12 (1), pp. 53–58.

Barak, Benny; Guiot, Denis; Mathur, Anil; Zhang, Yong; Lee, Keun (2011): An Empirical Assessment of Cross-Cultural Age Self-Construal Measurement. Evidence from Three Countries. In *Psychology & Marketing* 28 (5), pp. 479–495.

Barak, Benny; Schiffman, Leon G. (1981): Cognitive Age. A Nonchronological Age Variable. In *Advances in Consumer Research* 8 (1), pp. 602–606.

Baron, Reuben M.; Kenny, David A. (1986): The Moderator-Mediator Variable Distinction in Social Psychological Research. Conceptual, Strategic, and Statistical Considerations. In *Journal of Personality and Social Psychology* 51 (6), pp. 1173–1182.

Baumbach, Wenke; Schmidle, Michael (2008): Identifikation von Lead Usern in Online-Portalen am Beispiel von www.seniorenportal.de. Freising: TU München, Fakultät für Wirtschaftswissenschaft, Brau- und Lebensmittelindustrie (Consumer Science, 17).

Baumgartner, Hans; Homburg, Christian (1996): Applications of Structural Equation Modeling in Marketing and Consumer Research. A Review. In *International Journal of Research in Marketing* 13 (2), pp. 139–161.

Bearden, William O.; Netemeyer, Richard G.; Haws, Kelly L. (2011): Handbook of Marketing Scales. Multi-Item Measures for Marketing and Consumer Behavior Research. 3rd ed. Los Angeles, CA: Sage Publications.

Bearden, William O.; Netemeyer, Richard G.; Teel, Jesse E. (1989): Measurement of Consumer Susceptibility to Interpersonal Influence. In *Journal of Consumer Research* 15 (4), pp. 473–481.

Becker, Henk (2000): Discontinuous Change and Generational Contracts. In Sara Arber, Claudine Attias-Donfut (Eds.): The Myth of Generational Conflict. The Family and State in Ageing Societies. London, New York: Routledge, pp. 114–132.

Bengtson, Vern L.; Allen, Katherine R. (1993): The Life Course Perspective Applied to Families Over Time. In Pauline Boss, William J. Doherty, Ralph LaRossa, Walter R. Schumm, Suzanne K. Steinmetz (Eds.): Sourcebook of Family Theories and Methods. A Contextual Approach. New York, NY: Plenum Press, pp. 469–498.

Bengtson, Vern L.; Silverstein, Merril; Putney, Norella M.; Gans, Daphna (Eds.) (2009a): Handbook of Theories of Aging. 2nd ed. New York, NY: Springer.

Bengtson, Vern L.; Silverstein, Merril; Putney, Norella M.; Gans, Daphna (2009b): Preface. In Vern L. Bengtson, Merril Silverstein, Norella M. Putney, Daphna Gans (Eds.): Handbook of Theories of Aging. 2nd ed. New York, NY: Springer, pp. xxi–xxiii.

Bergmann, Bärbel; Prescher, Claudia; Eisfeldt, Doreen (2006): Alterstrends der Innovationstätigkeit bei Erwerbstätigen. In *Arbeit* 15 (1), pp. 18–28.

BITKOM (2011): Netzgesellschaft. Eine repräsentative Untersuchung zur Mediennutzung und dem Informationsverhalten der Gesellschaft in Deutschland. With assistance of Marcel Bertsch, Nathalie Huth, Rainer Arenz. Edited by Telekom-munikation und neue Medien e.V BITKOM Bundesverband Informationswirtschaft. Berlin. Available online at www.bitkom.org/files/documents/bitkom_publikation_netzgesellschaft.pdf, checked on 30/01/2014.

Blau, Zena Smith (1973): Old Age in a Changing Society. New York, NY: New Viewpoints.

Bleyer, Tobias; Windel, Armin; Müller-Arnecke, Heiner (2009): Produkte für Ältere? Produkte für alle! 1st ed. Dortmund: Bundesanstalt für Arbeitsschutz und Arbeitsmedizin (BAuA).

Bliemel, Friedhelm; Eggert, Andreas; Fassott, Georg; Henseler, Jörg (2005): Die PLS-Pfadmodellierung. Mehr als eine Alternative zur Kovarianzstrukturanalyse. In Friedhelm Bliemel, Andreas Eggert, Georg Fassott, Jörg Henseler (Eds.): Handbuch PLS-Pfadmodellierung. Methode, Anwendung, Praxisbeispiele. Stuttgart: Schäffer-Poeschel, pp. 9–16.

Bloom, David E.; Canning, David; Sevilla, Jaypee (2003): The Demographic Dividend. A New Perspective on the Economic Consequences of Population Change. Santa Monica: RAND.

Bogers, Marcel; Afuah, Allan; Bastian, Bettina (2010): Users as Innovators. A Review, Critique, and Future Research Directions. In *Journal of Management* 36 (4), pp. 857–875.

Bortz, Jürgen; Döring, Nicola (2009): Forschungsmethoden und Evaluation. Für Human- und Sozialwissenschaftler. Mit 87 Tabellen. 4th ed. Heidelberg: Springer Medizin Verlag.

Bortz, Jürgen; Schuster, Christof (2010): Statistik für Human- und Sozialwissenschaftler. Mit 70 Abbildungen und 163 Tabellen. 7th ed. Berlin, Heidelberg: Springer.

Braun, Viktor; Herstatt, Cornelius (2007): Barriers to User Innovation. Moving towards a Paradigm of 'Licence to Innovate'? In *International Journal of Technology, Policy and Management* 7 (3), pp. 292–303.

Braun, Viktor; Herstatt, Cornelius (2009): User-Innovation. Barriers to Democratization and IP Licensing. New York, NY: Routledge (Routledge Studies in Innovation, Organization and Technology, 13).

Bryman, Alan (2008): Social Research Methods. 3rd ed. Oxford: Oxford University Press.

Bullinger, Angelika C.; Rass, Matthias; Adamczyk, Sabrina (2011): Using Innovation Contests to Master Challenges of Demographic Change. Insights from Research and Practice. In Sven Kunisch, Stephan A. Boehm, Michael Boppel (Eds.): From Grey to Silver. Managing the Demographic Change Successfully. Berlin: Springer, pp. 163–174.

Bundesministerium für Wirtschaft und Technologie (BMWi), Deutschland (Ed.) (2010): Der Campingmarkt in Deutschland 2009/2010. Endbericht. Berlin (Studien, 587).

Bünstorf, Guido (2003): Designing Clunkers. Demand-side Innovation and the Early History of the Mountain Bike. In John Stan Metcalfe, Uwe Cantner (Eds.): Change, Transformation, and Development. With 71 Figures and 61 Tables. Heidelberg, New York: Physica-Verlag, pp. 53–70.

Campbell, Donald T.; Fiske, Donald W. (1959): Convergent and Discriminant Validation by the Multitrait-Multimethod Matrix. In *Psychological Bulletin* 56 (2), pp. 81-105.

Campinginfo.org (Ed.) (2012): The History of Camping. History of Recreational Camping in Britain and Elsewhere. Available online at www.campinginfo.org/family-camping/the-history-of-camping, updated in 2012, checked on 08/03/2013.

Caravaning Industrie Verband e.V. (CIVD) (Ed.) (2010): Soziodemografische Daten. Available online at www.civd.de/caravaning/marktzahlen/touristik/soziodemografische-daten/, updated on 22/11/2010, checked on 26/02/2013.

Caravaning Industrie Verband e.V. (CIVD) (Ed.) (2012): Freizeitfahrzeuge weltweit. Available online at www.civd.de/caravaning/marktzahlen/weltweit-freizeitfahrzeuge/, updated on 15/04/2012, checked on 26/02/2012.

Caravaning Industrie Verband e.V. (CIVD) (Ed.) (2013a): Deutsche Caravaningbranche setzt neue Höchstmarke. Erstmals mehr als 6 Milliarden Euro Gesamtumsatz. Available online at www.civd.de/caravaning/presse/pressemeldungen/2013/14-januar-2013-2/, updated on 10/01/2013, checked on 26/02/2013.

Caravaning Industrie Verband e.V. (CIVD) (Ed.) (2013b): Zahlen und Fakten Januar 2013. Available online at www.civd.de/fileadmin/civd/pdf/hintergrundinformationen/ Fact_sheet_Januar_2013.pdf, updated on 14/01/2013, checked on 26/02/2013.

Casper, Constance; Reichert, Irmingard (2008): Lead-User-Identifikation in virtuellen Communities am Beispiel des Frosta-Blogs. Freising: TU München, Fakultät für Wirtschaftswissenschaft, Brau- und Lebensmittelindustrie (Consumer Science, 19).

Charness, Neil; Krampe, Ralf T. (2008): Expertise and Knowledge. In Scott M. Hofer, Duane Francis Alwin (Eds.): Handbook of Cognitive Aging. Interdisciplinary Perspectives. Los Angeles, CA: Sage Publications, pp. 244–258.

Chin, Wynne W. (1998a): Issues and Opinion on Structural Equation Modeling. In *MIS Quarterly* 22 (1), pp. vii–xvi.

Chin, Wynne W. (1998b): The Partial Least Squares Approach to Structural Equation Modeling. In George A. Marcoulides (Ed.): Modern Methods for Business Research. Mahwah, N.J: Lawrence Erlbaum, pp. 295–336.

Chin, Wynne W. (2010): Bootstrap Cross-Validation Indices for PLS Path Model Assessment. In Vincenzo Esposito Vinzi, Wynne W. Chin, Jörg Henseler, Huiwen Wang (Eds.): Handbook of Partial Least Squares. Concepts, Methods and Applications. Berlin: Springer, pp. 83–97.

Chin, Wynne W.; Dibbern, Jens (2010): An Introduction to a Permutation Based Procedure for Multi-Group PLS Analysis. Results of Tests of Differences on Simulated Data and a Cross Cultural Analysis of the Sourcing of Information System Services Between Germany and the USA. In Vincenzo Esposito Vinzi, Wynne W. Chin, Jörg Henseler, Huiwen Wang (Eds.): Handbook of Partial Least Squares. Concepts, Methods and Applications. Berlin: Springer, pp. 171–193.

Chin, Wynne W.; Marcolin, Barbara L.; Newsted, Peter R. (2003): A Partial Least Squares Latent Variable Modeling Approach for Measuring Interaction Effects. Results from a Monte Carlo Simulation Study and an Electronic-Mail Emotion / Adoption Study. In *Information Systems Research* 14 (2), pp. 189–217.

Chin, Wynne W.; Newsted, Peter R. (1999): Structural Equation Modeling Analysis with Small Samples Using Partial Least Squares. In Rick H. Hoyle (Ed.): Statistical Strategies for Small Sample Research. 2nd ed. Thousand Oaks, CA: Sage Publications, pp. 307–342.

Churchill Jr., Gilbert A. (1979): A Paradigm for Developing Better Measures of Marketing Constructs. In *Journal of Marketing Research (JMR)* 16 (1), pp. 64–73.

Cierpicki, Steven; Wright, Malcolm; Sharp, Byron (2000): Managers' Knowledge of Marketing Principles. The Case of New Product Development. In *Journal of Empirical Generalisations in Marketing Science* 5 (3), pp. 771–790.

Claßen, Katrin (2012): Zur Psychologie von Technikakzeptanz im höheren Lebensalter. Die Rolle von Technikgenerationen. Dissertation. Ruprecht-Karls-Universität Heidelberg, Heidelberg. Psychologisches Insitut.

Cleaver, Megan; Muller, Thomas E. (2002): I Want to Pretend I'm Eleven Years Younger. Subjective Age and Seniors' Motives for Vacation Travel. In *Social Indicators Research* 60 (1), pp. 227–241.

Cohen, Jacob (1988): Statistical Power Analysis for the Behavioral Sciences. 2nd ed. Hillsdale, NJ: Erlbaum.

Cohen, Wesley M.; Levinthal, Daniel A. (1990): Absorptive Capacity. A New Perspective on Learning and Innovation. In *Administrative Science Quarterly* 35 (1), pp. 128–152.

Cole, Jason C. (2008): How to Deal With Missing Data. Conceptual Overview and Details for Implementing Two Modern Methods. In Jason W. Osborne (Ed.): Best Practices in Quantitative Methods. Thousand Oaks, CAS: Sage Publications, pp. 214–238.

Coulmas, Florian (2008): Looking at the Bright Side of Things. In Florian Kohlbacher, Cornelius Herstatt (Eds.): The Silver Market Phenomenon. Business Opportunities in an Era of Demographic Change. 1st ed. Berlin, Heidelberg: Springer, pp. v–vi.

Crawford, C. Merle (1977): Marketing Research and the New Product Failure Rate. In *Journal of Marketing* 41 (2), pp. 51–61.

Crown, William H. (1985): Some Thoughts on Reformulating the Dependency Ratio. In *The Gerontologist* 25 (2), pp. 166–171.

Dannefer, Dale (1987): Aging as Intracohort Differentiation. Accentuation, the Matthew Effect, and the Life Course. In *Sociological Forum* 2 (2), pp. 211-236.

Davenport, Thomas H.; Prusak, Laurence (1998): Working Knowledge. How Organizations Manage What They Know. Boston, MA: Harvard Business School Press.

Der Deutsche Camping-Club e.V. (Ed.) (2012): Deutschlands Camping-Fachverband. Der Deutsche Camping-Club e.V. 60 Jahre Einsatz für die Camper. Available online at www.camping-club.de/geschichte.0.html?&L=lvqmgldavew, updated in 2012, checked on 08/03/2013.

Deutsche Bundesbank (2013): Private Haushalte und ihre Finanzen. Tabellenanhang zur Pressenotiz vom 21.3.2013. PHF 2010/2011 - Datenstand: 2/2013. Available online at www.bundesbank.de/Redaktion/DE/Pressemitteilungen/BBK/2013/ 2013_03_21_phf.html, updated on 21/03/2013, checked on 22/06/2013.

Diamantopoulos, Adamantios (2006): The Error Term in Formative Measurement Models. Interpretation and Modeling Implications. In *Journal of Modelling in Management* 1 (1), pp. 7–17.

Diamantopoulos, Adamantios; Winklhofer, Heidi M. (2001): Index Construction with Formative Indicators. An Alternative to Scale Development. In *Journal of Marketing Research* 38 (2), pp. 269–277.

Dillman, Don A.; Smyth, Jolene D.; Christian, Leah Melani (2009): Internet, Mail, and Mixed-Mode Surveys. The Tailored Design Method. 3rd ed. Hoboken, NJ: Wiley & Sons.

Disney, Richard; Johnson, Paul (2001): Pension Systems and Retirement Incomes across OECD Countries. Cheltenham: Elgar.

Dömötör, Rudolf; Franke, Nikolaus; Hienerth, Christoph (2007): What a Difference a DV Makes … The Impact of Conceptualizing the Dependent Variable in Innovation Success Factor Studies. In *Zeitschrift für Betriebswirtschaft* 2, pp. 23–46.

Dychtwald, Ken; Flower, Joe (1990): Age Wave. The Challenges and Opportunities of an Aging America. New York, NY: Bantam Books.

Dziuban, Charles D.; Shirkey, Edwin C. (1974): When Is a Correlation Matrix Appropriate for Factor Analysis? Some Decision Rules. In *Psychological Bulletin* 81 (6), pp. 358–361.

Eisfeldt, Doreen (2009): Innovatives Arbeitsverhalten Erwerbstätiger. Bestandsaufnahme und wissensbasierte Ansatzpunkte zur Förderung innovativen Arbeitsverhaltens. Dissertation. Technische Universität Dresden, Dresden. Allgemeine Psychologie, Biopsychologie und Methoden der Psychologie.

Enomoto, Nozomi (2011): Changing Consumer Values and Behavior in Japan. Adaptation of Keio Department Store, Shinjuku. In Florian Kohlbacher, Cornelius Herstatt (Eds.): The Silver Market Phenomenon. Marketing and Innovation in the Aging Society. 2nd ed. Berlin, Heidelberg: Springer, pp. 175–193.

Enos, John Lawrence (1962): Petroleum Progress and Profits. A History of Process Innovation. Boston, Boston, MA: MIT Press.

Ernst, Holger; Soll, Jan Henrik; Spann, Martin (2004): Möglichkeiten der Lead-User-Identifikation in Online-Medien. In Cornelius Herstatt, Jan G. Sander (Eds.): Produktentwicklung mit virtuellen Communities. Kundenwünsche erfahren und Innovationen realisieren. 1st ed. Wiesbaden: Gabler, pp. 121–140.

Esposito Vinzi, Vincenzo; Trinchera, Laura; Amato, Silvano (2010): PLS Path Modeling. From Foundations to Recent Developments and Open Issues for Model Assessment and Improvement. In Vincenzo Esposito Vinzi, Wynne W. Chin, Jörg Henseler, Huiwen Wang (Eds.): Handbook of Partial Least Squares. Concepts, Methods and Applications. Berlin: Springer, pp. 47–82.

European Caravan Federation (Ed.) (2012a): Europe: Motor Caravans in Use. Registered 31.12.2011. Available online at www.e-c-f.org/fileadmin/templates/ 4825/images/statistics/E.7.2_2011.pdf, updated on 20/07/2012, checked on 26/02/2013.

European Caravan Federation (Ed.) (2012b): Europe: Touring Caravans in Use. Registered 31.12.2011. Available online at www.e-c-f.org/fileadmin/templates/ 4825/images/statistics/E.7.1_2011.pdf, updated on 20/07/2012, checked on 26/02/2013.

Fédération Internationale de Camping, Caravanning et Autocaravaning (Ed.) (2013): Presentation. Available online at http://ficc.org/txt.php?t=1, updated in 2013, checked on 12/03/2013.

Feld, Bradley A. (1990): The Changing Role of the User in the Development of Application Software. Cambridge, MA: Massachusetts Institute of Technology (Working Paper. Alfred P. Sloan School of Management, 3152-90-BPS).

Fent, Thomas; Mahlberg, Bernhard; Prskawetz, Alexia (2008): Demographic Change and Economic Growth. In Florian Kohlbacher, Cornelius Herstatt (Eds.): The Silver Market Phenomenon. Business Opportunities in an Era of Demographic Change. 1st ed. Berlin, Heidelberg: Springer, pp. 3–16.

Fichter, Klaus (2005): Nachhaltige Nutzerintegration im Innovationsprozess. In Klaus Fichter, Niko Paech, Reinhard Pfriem (Eds.): Nachhaltige Zukunftsmärkte. Orientierungen für unternehmerische Innovationsprozesse im 21. Jahrhundert. Marburg: Metropolis-Verlag (Theorie der Unternehmung, 29), pp. 351–370.

Fisk, Arthur D.; Rogers, Wendy A.; Charness, Neil; Czaja, Sara J.; Sharit, Joseph (2009): Designing for Older Adults. Principles and Creative Human Factors Approaches. 2nd ed. Boca Raton: CRC Press/Taylor & Francis (Human Factors & Aging Series).

Fitzen, Lena Katharina (2011): Kooperatives Distributionsmanagement und Distributionserfolg. Eine empirische Analyse aus Sicht des Automobilhandels. 1st ed. Lohmar, Köln: Josef Eul Verlag.

Flowers, Stephen; Hippel, Eric von; Jong, Jeroen P. J. de; Sinozic, Tanja (2010): Measuring User Innovation in the UK. The Importance of Product Creation by Users. Edited by NESTA. London. Available online at www.nesta.org.uk/publications/reports/assets/features/measuring_user_innovation_in_the_uk, checked on 14/05/2013.

Fornell, Claes; Bookstein, Fred L. (1982): Two Structural Equation Models. LISREL and PLS Applied to Consumer Exit-Voice Theory. In *Journal of Marketing Research* 19 (4), pp. 440–452.

Fornell, Claes; Larcker, David F. (1981): Evaluating Structural Equation Models with Unobservable Variables and Measurement Error. In *Journal of Marketing Research* 18 (1), pp. 39–50.

Foxall, Gordon R. (1995): Cognitive Styles of Consumer Initiators. In *Technovation* 15 (5), pp. 269–288.

Franke, Nikolaus; Hippel, Eric von (2003): Satisfying Heterogeneous User Needs via Innovation Toolkits. The Case of Apache Security Software. In *Research Policy* 32 (7), pp. 1199–1215.

Franke, Nikolaus; Hippel, Eric von; Schreier, Martin (2006): Finding Commercially Attractive User Innovations. A Test of Lead-User Theory. In *Journal of Product Innovation Management* 23 (4), pp. 301–315.

Franke, Nikolaus; Shah, Sonali (2003): How Communities Support Innovative Activities. An Exploration of Assistance and Sharing among End-Users. In *Research Policy* 32 (1), pp. 157–178.

Freeman, Christopher (1968): Chemical Process Plant. Innovation and the World Market. In *National Institute of Economic and Social Research*, pp. 29–51.

Frese, Michael; Kring, Wolfgang; Soose, Andrea; Zempel, Jeannette (1996): Personal Initiative at Work. Differences between East and West Germany. In *Academy of Management Journal* 39 (1), pp. 37–63.

Fuchs, Christoph; Diamantopoulos, Adamantios (2009): Using Single-Item Measures for Construct Measurement in Management Research. Conceptual Issues and Application Guidelines. In *Die Betriebswirtschaft* 69 (2), pp. 195–210.

Füller, Johann; Jawecki, Gregor; Mühlbacher, Hans (2007): Innovation Creation by Online Basketball Communities. In *Journal of Business Research* 60 (1), pp. 60–71.

Gassmann, Oliver; Reepmeyer, Gerrit (2006): Wachstumsmarkt Alter. Innovationen für die Zielgruppe 50+. München, Wien: Hanser.

Gassmann, Oliver; Reepmeyer, Gerrit (2008): Universal Design - Innovations for All Ages. In Florian Kohlbacher, Cornelius Herstatt (Eds.): The Silver Market Phenomenon. Business Opportunities in an Era of Demographic Change. 1st ed. Berlin, Heidelberg: Springer, pp. 125–140.

Geisser, Seymour (1975): The Predictive Sample Reuse Method with Applications. In *Journal of the American Statistical Association* 70 (350), pp. 320–328.

Goldsmith, Ronald E.; Hofacker, Charles F. (1991): Measuring Consumer Innovativeness. In *Journal of the Academy of Marketing Science* 19 (3), pp. 209–221.

Gourville, John T. (2005): The Curse of Innovation. Why Innovative New Products Fail. Report No. 05-117. In *MSI Reports: Working Paper Series* (4), pp. 3–23.

Green, Robert T.; Langeard, Eric (1975): A Cross-National Comparison of Consumer Habits and Innovator Characteristics. In *International Executive* 17 (3), pp. 5–7.

Griffin, Abbie (1997): PDMA Research on New Product Development Practices. Updating Trends and Benchmarking Best Practices. In *Journal of Product Innovation Management* 14 (6), pp. 429–458.

Grossmann, Igor; Na, Jinkyung; Varnum, Michael E. W.; Park, Denise C.; Kitayama, Shinobu; Nisbett, Richard E. (2010): Reasoning about Social Conflicts Improves into Old Age. In *Proceedings of the National Academy of Sciences* 107 (16), pp. 7246–7250.

Gruca, Thomas S.; Schewe, Charles D. (1992): Researching Older Consumers. In *Marketing Research* 4 (3), pp. 18–24.

Gwinner, Kevin P.; Stephens, Nancy (2001): Testing the Implied Mediational Role of Cognitive Age. In *Psychology & Marketing* 18 (10), pp. 1031–1048.

Hage, Jerald (1994): Formal Theory in Sociology. Opportunity or Pitfall? Albany, NY: State University of New York Press (SUNY Series on the New Inequalities).

Haigh, Ruth (1993): The Ageing Process. A Challenge for Design. In *Applied Ergonomics* 24 (1), pp. 9–14.

Hair, Joe F.; Ringle, Christian M.; Sarstedt, Marko (2011): PLS-SEM: Indeed a Silver Bullet. In *Journal of Marketing Theory & Practice* 19 (2), pp. 139–152.

Hair, Joe F.; Sarstedt, Marko; Ringle, Christian M.; Mena, Jeannette (2012): An Assessment of the Use of Partial Least Squares Structural Equation Modeling in Marketing Research. In *Journal of the Academy of Marketing Science* 40 (3), pp. 414–433.

Hair, Joseph F.; Black, William C.; Babin, Barry J.; Anderson, Rolph E. (2008): Multivariate Data Analysis. A Global Perspective. 7th ed. Upper Saddle River, NJ: Pearson.

Hair, Joseph F.; Hult, G. Tomas M.; Ringle, Christian M.; Sarstedt, Marko (2013): A Primer on Partial Least Squares Structural Equation Modeling (PLS-SEM). Thousand Oaks, CA: Sage Publications.

Harhoff, Dietmar; Henkel, Joachim; Hippel, Eric von (2003): Profiting from Voluntary Information Spillovers. How Users Benefit by Freely Revealing their Innovations. In *Research Policy* 32 (10), pp. 1753–1769.

Harzing, Anne-Wil; Wal, Ron van der (2008): Google Scholar as a New Source for Citation Analysis. In *Ethics in Science and Environmental Politics* 8, pp. 61–73.

Hauser, Heinrich (1935): Fahrten und Abenteuer im Wohnwagen. Dresden: Reißner.

Haustein, Thomas; Mischke, Johanna (2011): Im Blickpunkt: Ältere Menschen in Deutschland und der EU. Wiesbaden: Statistisches Bundesamt.

Hedrick-Wong, Yuwa (2007): The Glittering Silver Market. The Rise of Elderly Consumers in Asia. Singapore: Wiley.

Helminen, Pia (2008): Disabled Persons as Lead Users for Silver Market Customers. In Florian Kohlbacher, Cornelius Herstatt (Eds.): The Silver Market Phenomenon. Business Opportunities in an Era of Demographic Change. 1st ed. Berlin, Heidelberg: Springer, pp. 85–102.

Henderson, Kenneth V.; Goldsmith, Ronald E.; Flynn, Leisa R. (1995): Demographic Characteristics of Subjective Age. In *Journal of Social Psychology* 135 (4), pp. 447–457.

Henseler, Jörg; Chin, Wynne W. (2010): A Comparison of Approaches for the Analysis of Interaction Effects between Latent Variables Using Partial Least Squares Path Modeling. In *Structural Equation Modeling* 17 (1), pp. 82–109.

Henseler, Jörg; Fassott, Georg (2010): Testing Moderating Effects in PLS Path Models. An Illustration of Available Procedures. In Vincenzo Esposito Vinzi, Wynne W. Chin, Jörg Henseler, Huiwen Wang (Eds.): Handbook of Partial Least Squares. Concepts, Methods and Applications. Berlin: Springer, pp. 713–735.

Henseler, Jörg; Fassott, Georg; Dijkstra, Theo K.; Wilson, Bradley (2012a): Analysing Quadratic Effects of Formative Constructs by Means of Variance-based Structural Equation Modelling. In *European Journal of Information Systems* 21 (1), pp. 99–112.

Henseler, Jörg; Ringle, Christian M.; Sarstedt, Marko (2012b): Using Partial Least Squares Path Modeling in International Advertising Research. Basic Concepts and Recent Issues. In Shintaro Okazaki (Ed.): Handbook of Research on International Advertising. Cheltenham: Edward Elgar Publishing, pp. 252–276.

Henseler, Jörg; Ringle, Christian M.; Sinkovics, Rudolf R. (2009): The Use of Partial Least Squares Path Modeling in International Marketing. In *Advances in International Marketing* 20, pp. 277–319.

Henseler, Jörg; Sarstedt, Marko (2013): Goodness-of-Fit Indices for Partial Least Squares Path Modeling. In *Computational Statistics* 28 (2), pp. 565-580.

Herstatt, Cornelius; Hippel, Eric von (1992): From Experience: Developing New Product Concepts via the Lead User Method. A Case Study in a "Low-Tech" Field. In *Journal of Product Innovation Management* 9 (3), pp. 213–221.

Herstatt, Cornelius; Lüthje, Christian; Lettl, Christopher (2001): Innovation Search Fields with Lead Users. Hamburg (Working Paper // Technologie- und Innovationsmanagement). Available online at http://doku.b.tu-harburg.de/volltexte/2006/159/, checked on 26/02/2014.

Herstatt, Cornelius; Sander, Jan G. (Eds.) (2004): Produktentwicklung mit virtuellen Communities. Kundenwünsche erfahren und Innovationen realisieren. 1st ed. Wiesbaden: Gabler.

Hienerth, Christoph (2006): The Commercialization of User Innovations. The Development of the Rodeo Kayak Industry. In *R&D Management* 36 (3), pp. 273–294.

Hierhammer, Alfons (1997): Die Caravan- und Motorcaravanbranche in Deutschland. Phasen der langfristigen Entwicklung, Situation heute, Tendenzen für die Zukunft. Stuttgart: CDS-Verlag.

Hippel, Eric von (1976a): Dominant Role of User in Semiconductor and Electronic Subassembly Process Innovation. Edited by MIT Sloan School of Management. Cambridge, MA.

Hippel, Eric von (1976b): The Dominant Role of Users in the Scientific Instrument Innovation Process. In *Research Policy* 5 (3), pp. 212–239.

Hippel, Eric von (1986): Lead Users. A Source of Novel Product Concepts. In *Management Science* 32 (7), pp. 791–805.

Hippel, Eric von (1988): The Sources of Innovation. New York, NY: Oxford University Press.

Hippel, Eric von (1994): "Sticky Information" and the Locus of Problem Solving. Implications for Innovation. In *Management Science* 40 (4), pp. 429–439.

Hippel, Eric von (2005a): Democratizing Innovation. Cambridge, MA: MIT Press.

Hippel, Eric von (2005b): Democratizing Innovation. The Evolving Phenomenon of User Innovation. In *Journal für Betriebswirtschaft* 55 (1), pp. 63-78.

Hippel, Eric von (2007): Horizontal Innovation Networks - By and for Users. In *Industrial and Corporate Change* 16 (2), pp. 293–315.

Hippel, Eric von; Jong, Jeroen P. J. de; Flowers, Stephen (2012): Comparing Business and Household Sector Innovation in Consumer Products. Findings from a Representative Study in the United Kingdom. In *Management Science* 58 (9), pp. 1669–1681.

Hippel, Eric von; Krogh, Georg von (2003): Open Source Software and the "Private-Collective" Innovation Model. Issues for Organization Science. In *Organization Science* 14 (2), pp. 209–223.

Hippel, Eric von; Ogawa, Susumu; Jong, Jeroen P. J. de (2011): The Age of the Consumer-Innovator. In *MIT Sloan Management Review* 53 (1), pp. 27–35.

Hippel, Eric von; Thomke, Stefan; Sonnack, Mary (1999): Creating Breakthroughs at 3M. In *Harvard Business Review* 77 (5), pp. 47–57.

Hofer, Scott M.; Alwin, Duane Francis (Eds.) (2008): Handbook of Cognitive Aging. Interdisciplinary Perspectives. Los Angeles, CA: Sage Publications.

Hoffmann, Stefan; Soyez, Katja (2010): A Cognitive Model to Predict Domain-Specific Consumer Innovativeness. In *Journal of Business Research* 63 (7), pp. 778–785.

Hoisl, Karin (2007): A Closer Look at Inventive Output. The Role of Age and Career Paths. Edited by Ludwig-Maximilians-Universität München. Institute for Innovation Research, Technology Management and Entrepreneurship. München.

Holding, Thomas Hiram (1898): Cycle and Camp. London: Ward, Lock & Co.

Holding, Thomas Hiram (1908): The Camper's Handbook. Special Contributions by the Lady Arthur Grosvenor; Mrs. Horsfield; Matthew Arnold; G.D. Matthews; R.J. Mecredy. [With illustrations.]. Simpkin, Marshall & Co: London.

Holford, Theodore R. (1983): The Estimation of Age, Period and Cohort Effects for Vital Rates. In *Biometrics* 39 (2), pp. 311–324.

Homburg, Christian; Klarmann, Martin (2006): Die Kausalanalyse in der empirischen betriebswirtschaftlichen Forschung. Problemfelder und Anwendungsempfehlungen. In *Die Betriebswirtschaft* 66 (6), pp. 727–748.

Hooyman, Nancy R.; Kiyak, H. Asuman (2011): Social Gerontology. A Multidisciplinary Perspective. 9th ed. Boston, MA: Allyn & Bacon.

Horn, John L.; Cattell, Raymond B. (1967): Age Differences in Fluid and Crystallized Intelligence. In *Acta Psychologica* 26, pp. 107–129.

Huber, Frank; Herrmann, Andreas; Meyer, Frederik; Vogel, Johannes; Vollhardt, Kai (2007): Kausalmodellierung mit Partial Least Squares. Eine anwendungsorientierte Einführung. Wiesbaden: Betriebswirtschaftlicher Verlag Dr. Th. Gabler | GWV Fachverlage GmbH Wiesbaden.

Hubley, Anita M.; Hultsch, David F. (1994): The Relationship of Personality Trait Variables to Subjective Age Identity in Older Adults. In *Research on Aging* 16 (4), pp. 415–439.

Hubley, Anita M.; Hultsch, David F. (1996): Subjective Age and Traits. In *Research on Aging* 18 (4), pp. 494–496.

Iffländer, Klaus; Levsen, Nils; Lorscheid, Iris; Pakur, Sandra; Wellner, Konstantin; Herstatt, Cornelius; Lüthje, Christian; Meyer, Matthias; Ringle, Christian M. (2012): Innoage. Innovation and Product Development for Aging Users. Edited by Technische Universität Hamburg-Harburg (Management@TUHH, 6).

Im, Subin; Bayus, Barry L.; Mason, Charlotte H. (2003): An Empirical Study of Innate Consumer Innovativeness, Personal Characteristics, and New-Product Adoption Behavior. In *Journal of the Academy of Marketing Science* 31 (1), pp. 61–73.

Janzik, Lars (2012): Motivanalyse zu Anwenderinnovationen in Online-Communities. 1st ed. Wiesbaden: Gabler (Forschungs-/Entwicklungs-/Innovations-Management).

Jarvis, Cheryl Burke; MacKenzie, Scott B.; Podsakoff, Philip M.; Mick, David Glen; Bearden, William O. (2003): A Critical Review of Construct Indicators and Measurement Model Misspecification in Marketing and Consumer Research. In *Journal of Consumer Research* 30 (2), pp. 199–218.

Jeppesen, Lars Bo; Frederiksen, Lars (2006): Why Do Users Contribute to Firm-Hosted User Communities? The Case of Computer-Controlled Music Instruments. In *Organization Science* 17 (1), pp. 45–63.

Jeppesen, Lars Bo; Laursen, Keld (2009): The Role of Lead Users in Knowledge Sharing. In *Research Policy* 38 (10), pp. 1582–1589.

Jiptner, Katharina; Bialkowski, Alexander; Chieng, Chiong Leong (2009): Lead-User-Identifikation in virtuellen Communities am Beispiel von www.biertest-online.de. Freising: TU München, Fakultät für Wirtschaftswissenschaft, Brau- und Lebensmittelindustrie.

Jong, Jeroen P. J. de; Hippel, Eric von (2009): Transfers of User Process Innovations to Process Equipment Producers. A Study of Dutch High-Tech Firms. In *Research Policy* 38 (7), pp. 1181–1191.

Kahle, Lynn R. (1983): Social Values and Social Change. Adaptation to Life in America. New York, NY: Praeger.

Kaiser, Henry F.; Rice, John (1974): Little Jiffy, Mark IV. In *Educational and Psychological Measurement* 34 (1), pp. 111–117.

Kastenbaum, R.; Derbin, V.; Sabatini, P.; Artt, S. (1972): "The Ages of Me". Toward Personal and Interpersonal Definitions of Functional Aging. In *International Journal of Aging and Human Development* 3 (2), pp. 197–211.

Kirton, Michael J. (1976): Adaptors and Innovators. A Description and Measure. In *Journal of Applied Psychology* 61 (5), pp. 622–629.

Kock, Ned; Chatelain-Jardon, Ruth; Carmona, Jesus (2008): An Experimental Study of Simulated Web-based Threats and Their Impact on Knowledge Communication Effectiveness. In *IEEE Transactions on Professional Communication* 51 (2), pp. 183–197.

Kohlbacher, Florian; Chéron, Emmanuel (2011): Understanding "Silver" Consumers through Cognitive Age, Health Condition, Financial Status, and Personal Values. Empirical Evidence from the World's Most Mature Market Japan. In *Journal of Consumer Behaviour*, pp. 179–188.

Kohlbacher, Florian; Gudorf, Pascal; Herstatt, Cornelius (2011a): Japan's Growing Silver Market. An Attractive Business Opportunity for Foreign Companies? In Sven Kunisch, Stephan A. Boehm, Michael Boppel (Eds.): From Grey to Silver. Managing the Demographic Change Successfully. Berlin: Springer, pp. 189–205.

Kohlbacher, Florian; Herstatt, Cornelius (2008a): Preface and Introduction. In Florian Kohlbacher, Cornelius Herstatt (Eds.): The Silver Market Phenomenon. Business Opportunities in an Era of Demographic Change. 1st ed. Berlin, Heidelberg: Springer, pp. xi–xxv.

Kohlbacher, Florian; Herstatt, Cornelius (Eds.) (2008b): The Silver Market Phenomenon. Business Opportunities in an Era of Demographic Change. 1st ed. Berlin, Heidelberg: Springer.

Kohlbacher, Florian; Herstatt, Cornelius (2011a): Preface and Introduction. In Florian Kohlbacher, Cornelius Herstatt (Eds.): The Silver Market Phenomenon. Marketing and Innovation in the Aging Society. 2nd ed. Berlin, Heidelberg: Springer, pp. v–xxii.

Kohlbacher, Florian; Herstatt, Cornelius (Eds.) (2011b): The Silver Market Phenomenon. Marketing and Innovation in the Aging Society. 2nd ed. Berlin, Heidelberg: Springer.

Kohlbacher, Florian; Herstatt, Cornelius; Levsen, Nils (in press): Golden Opportunities for Silver Innovation. How Demographic Changes Give Rise to New Entrepreneurial Opportunities for Meeting the Needs of Older People. In *Technovation*.

Kohlbacher, Florian; Herstatt, Cornelius; Schweisfurth, Tim (2011b): Product Development for the Silver Market. In Florian Kohlbacher, Cornelius Herstatt (Eds.): The Silver Market Phenomenon. Marketing and Innovation in the Aging Society. 2nd ed. Berlin, Heidelberg: Springer, pp. 3–13.

Kohli, Martin (1985): Die Institutionalisierung des Lebenslaufs. Historische Befunde und theoretische Argumente. In *Kölner Zeitschrift für Soziologie und Sozialpsychologie* 37 (1), pp. 1–29.

Kotler, Philip; Bliemel, Friedhelm (2006): Marketing-Management. Analyse, Planung und Verwirklichung. 10th ed. München: Pearson Studium.

Kratzer, Jan; Lettl, Christopher (2008): A Social Network Perspective of Lead Users and Creativity. An Empirical Study among Children. In *Creativity & Innovation Management* 17 (1), pp. 26–36.

Kunisch, Sven; Boehm, Stephan A.; Boppel, Michael (Eds.) (2011): From Grey to Silver. Managing the Demographic Change Successfully. Berlin: Springer.

Leeuw, E.D de (2005): To Mix or Not to Mix Data Collection Modes in Surveys. In *Journal of Official Statistics* 21 (5), pp. 233–255.

Lettl, Christopher; Gemünden, Hans Georg (2005): The Entrepreneurial Role of Innovative Users. In *Journal of Business & Industrial Marketing* 20 (7), pp. 339–346.

Lettl, Christopher; Herstatt, Cornelius; Gemünden, Hans Georg (2006): Learning from Users for Radical Innovation. In *International Journal of Technology Management* 33 (1), p. 5.

Lévesque, Moren; Minniti, Maria (2006): The Effect of Aging on Entrepreneurial Behavior. In *Journal of Business Venturing* 21 (2), pp. 177–194.

Levsen, Nils (2015): Lead Markets in Age-Based Innovations. Demographic Change and Internationally Successful Innovations. Wiesbaden: Springer Gabler.

Leyhausen, Frank; Vossen, Alexander (2011): We Could Have Known Better. Consumer-Oriented Marketing in Germany's Ageing Market. In Sven Kunisch, Stephan A. Boehm, Michael Boppel (Eds.): From Grey to Silver. Managing the Demographic Change Successfully. Berlin: Springer, pp. 175–184.

Liang, Huigang; Saraf, Nilesh; Qing Hu; Xue, Yajiong (2007): Assimilation of Enterprise Systems. The Effect of Institutional Pressures and the Mediating Role of Top Management. In *MIS Quarterly* 31 (1), pp. 59–87.

Lilien, Gary L.; Morrison, Pamela D.; Searls, Kathleen; Sonnack, Mary; Hippel, Eric von (2002): Performance Assessment of the Lead User Idea-Generation Process for New Product Development. In *Management Science* 48 (8), pp. 1042–1059.

Ludwig, Frederic C.; Smoke, Mary E. (1980): The Measurement of Biological Age. In *Experimental Aging Research* 6 (6), pp. 497–522.

Lumpkin, James R.; Caballero, Marjorie J.; Chonko, Lawrence B. (1989): Direct Marketing, Direct Selling, and the Mature Consumer. A Research Study. New York, NY: Quorum Books.

Lüthje, Christian (2000): Kundenorientierung im Innovationsprozess. Eine Untersuchung der Kunden-Hersteller-Interaktion in Konsumgütermärkten. Dissertation. Wiesbaden: Deutscher Universitäts-Verlag (Betriebswirtschaftslehre für Technologie und Innovation, 33).

Lüthje, Christian (2003): Customers as Co-Inventors. An Empirical Analysis of the Antecedents of Customer-Driven Innovations in the Field of Medical Equipment. In European Marketing Academy (Ed.): Proceedings of the 32th EMAC Conference. EMAC Annual Conference. Glasgow, 20-23/05/2003.

Lüthje, Christian (2004): Characteristics of Innovating Users in a Consumer Goods Field. An Empirical Study of Sport-Related Product Consumers. In *Technovation* 24 (9), pp. 683–695.

Lüthje, Christian; Herstatt, Cornelius (2004): The Lead User Method. An Outline of Empirical Findings and Issues for Future Research. In *R&D Management* 34 (5), pp. 553–568.

Lüthje, Christian; Herstatt, Cornelius; Hippel, Eric von (2002): The Dominant Role of "Local" Information in User Innovation. The Case of Mountain Biking. MIT Working Paper 4377-02. Edited by MIT Sloan School of Management. Cambridge, MA.

Lüthje, Christian; Herstatt, Cornelius; Hippel, Eric von (2005): User-Innovators and "Local" Information. The Case of Mountain Biking. In *Research Policy* 34 (6), pp. 951–965.

MacKenzie, Scott B.; Podsakoff, Philip M.; Jarvis, Cheryl Burke (2005): The Problem of Measurement Model Misspecification in Behavioral and Organizational Research and Some Recommended Solutions. In *Journal of Applied Psychology* 90 (4), pp. 710–730.

MacKenzie, Scott B.; Podsakoff, Philip M.; Podsakoff, Nathan P. (2011): Construct Measurement and Validation Procedures in MIS and Behavioral Research. Integrating New and Existing Techniques. In *MIS Quarterly* 35 (2), pp. 293–334.

Malanowski, Norbert (2008): Matching Demand and Supply: Future Technologies for Active Ageing in Europe. In Florian Kohlbacher, Cornelius Herstatt (Eds.): The Silver Market Phenomenon. Business Opportunities in an Era of Demographic Change. 1st ed. Berlin, Heidelberg: Springer, pp. 41–53.

Mann, H. B.; Whitney, D. R. (1947): On a Test of Whether One of Two Random Variables is Stochastically Larger than the Other. In *Annals of Mathematical Statistics* 18 (1), pp. 50–60.

Mansfield, Edwin (1986): Patents and Innovation. An Empirical Study. In *Management Science* 32 (2), pp. 173–181.

Marchi, Gianluca; Giachetti, Claudio; Gennaro, Pamela de (2011): Extending Lead-User Theory to Online Brand Communities. The Case of the Community Ducati. In *Technovation* 31 (8), pp. 350–361.

Marcoulides, George A.; Saunders, Carol (2006): Editor's Comments. PLS: A Silver Bullet? In *MIS Quarterly* 30 (2), pp. iii–ix.

Mayer, John D.; Faber, Michael A.; Xu, Xiaoyan (2007): Seventy-five Years of Motivation Measures (1930–2005). A Descriptive Analysis. In *Motivation and Emotion* 31 (2), pp. 83-103.

Mayer, Karl Ulrich (1990): Lebensverläufe und sozialer Wandel. Anmerkungen zu einem Forschungsprogramm. In *Lebensverläufe und sozialer Wandel (Kölner Zeitschrift für Soziologie und Sozialpsychologie)* (Sonderheft 31), pp. 7–21.

Midgley, David F. (1977): Innovation and New Product Marketing. London: Croom Helm.

Midgley, David F.; Dowling, Grahame R. (1978): Innovativeness. The Concept and its Measurement. In *Journal of Consumer Research* 4 (4), pp. 229–242.

Midgley, David F.; Dowling, Grahame R. (1993): A Longitudinal Study of Product Form Innovation. The Interaction between Predispositions and Social Messages. In *Journal of Consumer Research* 19 (4), pp. 611–625.

Morrison, Margie; McMillan, Sally (2010): Oh, User, Who Art Thou. An Examination of Behaviors and Characteristics of Consumers in the Context of User Generated Content. In *American Academy of Advertising Conference Proceedings*, p. 77.

Morrison, Pamela D. (1996): Testing a Framework for the Adoption of Technological Innovations by Organizations and the Role of Leading Edge Users. Edited by The Pennsylvania State University. Institute for the Study of Business Markets (Working Papers, 17-1996).

Morrison, Pamela D.; Roberts, John H.; Hippel, Eric von (2000): Determinants of User Innovation and Innovation Sharing in a Local Market. In *Management Science* 46 (12), pp. 1513–1527.

Morrison, Pamela D.; Roberts, John H.; Midgley, David F. (1999): Towards a Finer Understanding of Lead Users. ISBM Report 15-1999. Edited by Institute for the Study of Business Markets. The Pennsylvania State University.

Morrison, Pamela D.; Roberts, John H.; Midgley, David F. (2004): The Nature of Lead Users and Measurement of Leading Edge Status. In *Research Policy* 33 (2), p. 351.

Moschis, George P. (1992a): GERONTOGRAPHICS. A Scientific Approach to Analyzing and Targeting the Mature Market. In *Journal of Services Marketing* 6 (3), pp. 17–26.

Moschis, George P. (1992b): Marketing to Older Consumers. A Handbook of Information for Strategy Development. Westport, Conn: Quorum Books.

NC State University, The Center for Universal Design (1997): The Principles of Universal Design. With assistance of Bettye Rose Connel, Mike Jones, Ron Mace, Jim Mueller, Abir Mullick, Elaine Ostroff et al. Available online at www.ncsu.edu/www/ncsu/design/sod5/cud/about_ud/udprinciplestext.htm, updated on 01/04/1997, checked on 30/01/2014.

Netemeyer, Richard G.; Bearden, William O.; Sharma, Subhash (2003): Scaling Procedures. Issues and Applications. Thousand Oaks, CA: Sage Publications.

Nonaka, Ikuji; Takeuchi, Hirotaka (1995): The Knowledge Creating Company. How Japanese Companies Create the Dynamics of Innovation. New York, NY: Oxford University Press.

Nunnally, Jum C.; Bernstein, Ira H. (1994): Psychometric Theory. 3rd ed. New York, NY: McGraw-Hill (McGraw-Hill Series in Psychology).

Oberg, Winston (1960): Age and Achievement – and the Technical Man. In *Personnel Psychology* 13 (3), pp. 245–259.

Office for National Statistics (2012): Family Spending. 2012 Edition. Edited by UK Statistics Authority. Available online at www.ons.gov.uk/ons/rel/family-spending/family-spending/family-spending-2012-edition/index.html, updated on 04/12/2012, checked on 21/06/2013.

Ogawa, Susumu (1998): Does Sticky Information Affect the Locus of Innovation? Evidence from the Japanese Convenience-Store Industry. In *Research Policy* 26 (7-8), pp. 777–790.

Ogawa, Susumu; Pongtanalert, Kritinee (2011): Visualizing Invisible Innovation Continent. Evidence from Global Consumer Innovation Surveys. In *SSRN Journal* (30/06/2011). Available only at http://ssrn.com/abstract=1876186, checked on 26/02/2014.

Ogawa, Susumu; Pongtanalert, Kritinee (2013): Exploring Characteristics and Motives of Consumer Innovators. In *Research Technology Management* 56 (3), pp. 41–48.

Ogawa, Takeo (2008): Changing Social Concepts of Age. Towards the Active Senior Citizen. In Florian Coulmas, Harald Conrad, Annette Schad-Seifert, Gabriele Vogt (Eds.): The Demographic Challenge. A Handbook about Japan. Leiden: Koninklijke Brill NV, pp. 145–161.

Olson, Erik L.; Bakke, Geir (2001): Implementing the Lead User Method in a High Technology Firm. A Longitudinal Study of Intentions versus Actions. In *Journal of Product Innovation Management* 18 (6), pp. 388–395.

Organisation for Economic Co-operation and Development (OECD); Statistical Office of the European Communities (Eurostat) (2005): Oslo Manual. Guidelines for Collecting and Interpreting Innovation Data. 3rd ed. Paris (The Measurement of Scientific and Technological Activities).

Osborne, Jason W.; Overbay, Amy (2008): Best Practices in Data Cleaning. How Outliers and "Fringeliers" Can Increase Error Rates and Decrease the Quality and Precision of Your Results. In Jason W. Osborne (Ed.): Best Practices in Quantitative Methods. Thousand Oaks, CA: Sage Publications, pp. 205–213.

Östlund, Britt (2011): Silver Age Innovators. A New Approach to Old Users. In Florian Kohlbacher, Cornelius Herstatt (Eds.): The Silver Market Phenomenon. Marketing and Innovation in the Aging Society. 2nd ed. Berlin, Heidelberg: Springer, pp. 15–26.

Outdoor Foundation (Ed.) (2012): 2012 American Camper Report. A Look Back and the Year Ahead. Available online at www.outdoorfoundation.org/pdf/research.camping.2012.pdf, checked on 26/02/2013.

Pettigrew, Simone (2008): Older Consumers' Customer Service Preferences. In Florian Kohlbacher, Cornelius Herstatt (Eds.): The Silver Market Phenomenon. Business Opportunities in an Era of Demographic Change. 1st ed. Berlin, Heidelberg: Springer, pp. 257–268.

Pfeiffer, Günter (2011): How Swisscom Copes with the Challenges of Demographic Change. In Sven Kunisch, Stephan A. Boehm, Michael Boppel (Eds.): From Grey to Silver. Managing the Demographic Change Successfully. Berlin: Springer, pp. 185–187.

Pirkl, James J. (2011): Transgenerational Design: A Heart Transplant for Housing. In Florian Kohlbacher, Cornelius Herstatt (Eds.): The Silver Market Phenomenon. Marketing and Innovation in the Aging Society. 2nd ed. Berlin, Heidelberg: Springer, pp. 117–131.

Plous, Scott (1993): The Psychology of Judgment and Decision Making. New York, NY: McGraw-Hill (McGraw-Hill Series in Social Psychology).

Plutzer, Eric; Berkman, Michael (2005): The Graying of America and Support for Funding the Nation's Schools. In *Public Opinion Quarterly* 69 (1), pp. 66–86.

Podsakoff, Philip M.; MacKenzie, Scott B.; Jeong-Yeon Lee; Podsakoff, Nathan P. (2003): Common Method Biases in Behavioral Research. A Critical Review of the Literature and Recommended Remedies. In *Journal of Applied Psychology* 88 (5), p. 879.

Podsakoff, Philip M.; Organ, Dennis W. (1986): Self-Reports in Organizational Research. Problems and Prospects. In *Journal of Management* 12 (4), pp. 531–544.

Polanyi, Michael (1958): Personal Knowledge. Towards a Post-Critical Philosophy. Chicago: University of Chicago Press.

Porst, Rolf (2000): Question Wording. Zur Formulierung von Fragebogen-Fragen. Edited by Zentrum für Umfragen, Methoden und Analysen. Mannheim (ZUMA How-to-Reihe, 2).

Preacher, Kristopher J; Hayes, Andrew F (2004): SPSS and SAS Procedures for Estimating Indirect Effects in Simple Mediation Models. In *Behavior Research Methods, Instruments, & Computers* 36 (4), pp. 717-731.

Raasch, Christina; Herstatt, Cornelius; Lock, Phillip (2008): The Dynamics of User Innovation. Drivers and Impediments of Innovation Activities. In *International Journal of Innovation Management* 12 (3), pp. 377–398.

Rammer, Christian; Aschhoff, Birgit; Crass, Dirk; Doherr, Thorsten; Hud, Martin; Köhler, Christian et al. (2013): Innovationsverhalten der deutschen Wirtschaft. Indikatorenbericht zur Innovationserhebung 2012. Edited by Zentrum für Europäische Wirtschaftsforschung GmbH (ZEW). Mannheim.

Randers, Ingrid; Mattiasson, Anne-Cathrine (2004): Autonomy and Integrity. Upholding Older Adult Patients' Dignity. In *Journal of Advanced Nursing* 45 (1), pp. 63–71.

Reinartz, Werner; Haenlein, Michael; Henseler, Jörg (2009): An Empirical Comparison of the Efficacy of Covariance-based and Variance-based SEM. In *International Journal of Research in Marketing* 26 (4), pp. 332–344.

Reinhardt, Andy (1998): Steve Jobs on Apple's Resurgence: "Not a One-Man Show". In *Bloomberg Businessweek* 1998, 12/05/1998.

Reinmöller, Patrick (2008): Service Innovation. Towards Designing New Business Models for Aging Societies. In Florian Kohlbacher, Cornelius Herstatt (Eds.): The Silver Market Phenomenon. Business Opportunities in an Era of Demographic Change. 1st ed. Berlin, Heidelberg: Springer, pp. 157–169.

Rigdon, Edward E.; Ringle, Christian M.; Sarstedt, Marko (2010): Structural Modeling of Heterogeneous Data with Partial Least Squares. In Naresh K. Malhotra (Ed.): Review of Marketing Research, Volume 7. Bingley: Emerald Group Publishing, pp. 255–296.

Rigdon, Edward E.; Schumacker, Randall E.; Wothke, Werner (1998): A Comparative Review of Interaction and Nonlinear Modeling. In George A. Marcoulides, Randall E. Schumacker (Eds.): Interaction and Nonlinear Effects in Structural Equation Modeling. Mahwah, N.J: Erlbaum.

Riggs, William; Hippel, Eric von (1994): Incentives to Innovate and the Sources of Innovation. The Case of Scientific Instruments. In *Research Policy* 23 (4), pp. 459–469.

Ringle, Christian M.; Wende, Sven; Will, S. (2005): SmartPLS 2.0 (M3) Beta. Hamburg. Available online at http://www.smartpls.de.

Robinson, John P.; Shaver, Phillip R.; Wrightsman, Lawrence S. (1991): Criteria for Scale Slection and Evaluation. In John P. Robinson, Phillip R. Shaver, Lawrence S. Wrightsman (Eds.): Measures of Social Psychological Attitudes: Academic Press, pp. 1–16.

Roehrich, Gilles (2004): Consumer Innovativeness. Concepts and Measurements. In *Journal of Business Research* 57 (6), p. 671.

Rogers, Everett Mitchell (1962): Diffusion of Innovations. New York, NY: Free Press.

Rogers, Everett Mitchell (2003): Diffusion of Innovations. 5th ed. New York, NY: Free Press.

Rogers, Everett Mitchell; Shoemaker, F. Floyd (1971): Communication of Innovations. A Cross-Cultural Approach. 2nd ed. New York, NY: Free Press.

Rossiter, John R. (2002): The C-OAR-SE Procedure for Scale Development in Marketing. In *International Journal of Research in Marketing* 19 (4), pp. 305–335.

Ryalls, Alan; Petri, Robert Lee: Camping. Edited by Encyclopedia Britannica (Encyclopedia Britannica Online). Available online at www.britannica.com/EBchecked/topic/91358/camping, checked on 08/03/2013.

Ryan, Richard M.; Deci, Edward L. (2000): Self-Determination Theory and the Facilitation of Intrinsic Motivation, Social Development, and Well-Being. In *American Psychologist* 55 (1), pp. 68–78.

Salthouse, Timothy A. (2009): When Does Age-related Cognitive Decline Begin? In *Neurobiology of Aging* 30 (4), pp. 507–514.

Sarstedt, Marko; Henseler, Jörg; Ringle, Christian M. (2011): Multigroup Analysis in Partial Least Squares (PLS) Path Modeling. Alternative Methods and Empirical Results. In: Measurement and Research Methods in International Marketing, Volume 22. Bingley: Emerald Group Publishing (Advances in International Marketing), pp. 195–218.

Saul, Darien Jay (15/01/2014): 3 Million Teens Leave Facebook in 3 Years. The 2014 Facebook Demographic Report. iStrategyLabs. Available online at http://istrategylabs.com/2014/01/3-million-teens-leave-facebook-in-3-years-the-2014-facebook-demographic-report/, checked on 16/01/2014.

Saup, Winfried (1993): Alter und Umwelt. Eine Einführung in die ökologische Gerontologie. Stuttgart: W. Kohlhammer.

Schaie, K. Warner; Willis, Sherry L. (Eds.) (2011): Handbook of the Psychology of Aging. 7th ed. Amsterdam: Academic Press.

Schapkin, Sergej A. (2012): Altersbezogene Änderungen kognitiver Fähigkeiten - kompensatorische Prozesse und physiologische Kosten. Forschung Projekt F 2152. Dortmund, Berlin, Dresden: Bundesanstalt für Arbeitsschutz und Arbeitsmedizin (BAuA).

Schewe, Charles D. (1991): Strategically Positioning Your Way into the Aging Marketplace. In *Business Horizons* 34 (3), pp. 59–66.

Schiffman, Leon G.; Sherman, Elaine (1991): Value Orientations of New-Age Elderly. The Coming of an Ageless Market. In *Journal of Business Research* 22 (2), pp. 187–194.

Schmidt-Ruhland, Karin; Knigge, Matthias (2008): Integration of the Elderly into the Design Process. In Florian Kohlbacher, Cornelius Herstatt (Eds.): The Silver Market Phenomenon. Business Opportunities in an Era of Demographic Change. 1st ed. Berlin, Heidelberg: Springer, pp. 103–124.

Schreier, Martin; Oberhauser, Stefan; Prügl, Reinhard (2007): Lead Users and the Adoption and Diffusion of New Products. Insights from Two Extreme Sports Communities. In *Marketing Letters* 18 (1-2), pp. 15-30.

Schreier, Martin; Prügl, Reinhard (2008): Extending Lead-User Theory. Antecedents and Consequences of Consumer's Lead Userness. In *Journal of Product Innovation Management* 25 (4), pp. 331–346.

Schuhmacher, Monika C.; Kuester, Sabine (2012): Identification of Lead User Characteristics Driving the Quality of Service Innovation Ideas. In *Creativity & Innovation Management* 21 (4), pp. 427–442.

Schumpeter, Joseph Alois (1934): [Theorie der wirtschaftlichen Entwicklung.] The Theory of Economic Development. An Inquiry into Profits, Capital, Credit, Interest, and the Business Cycle. Translated by Redvers Opie. Cambridge, MA: Harvard University Press.

Schumpeter, Joseph Alois (1942): Capitalism, Socialism and Democracy. New York, NY: Harper & Brothers.

Schweisfurth, Tim (2013): Embedded Lead Users Inside the Firm. How Innovative User Employees Contribute to the Corporate Product Innovation Process. Wiesbaden: Springer Gabler.

Schweisfurth, Tim; Raasch, Christina (2012): Lead Users as Firm Employees. How Are They Different and Why Does It Matter? In *SSRN Journal* (19/10/2011). Available online at http://ssrn.com/abstract=2164555, checked on 26/02/2014.

Selst, Mark van; Jolicoeur, Pierre (1994): A Solution to the Effect of Sample Size on Outlier Elimination. In *The Quarterly Journal of Experimental Psychology Section A* 47 (3), pp. 631–650.

Settersten, Richard A., Jr.; Mayer, Karl Ulrich (1997): The Measurement of Age, Age Structuring, and the Life Course. In *Annual Review of Sociology* 23 (1), pp. 233–261.

Shah, Sonali (2000): Sources and Patterns of Innovation in a Consumer Products Field. Innovations in Sporting Equipment. Sloan Working Paper #4105. Edited by MIT Sloan School of Management. Cambridge, MA.

Shapiro, Carl (2001): Navigating the Patent Thicket. Cross Licenses, Patent Pools, and Standard Setting. In Adam B. Jaffe, Joshua Lerner, Scott Stern (Eds.): Innovation Policy and the Economy. Volume 1. 1st ed. Cambridge, MA: MIT Press, pp. 119–150.

Shaw, Brian (1985): The Role of the Interaction between the User and the Manufacturer in Medical Equipment Innovation. In *R&D Management* 15 (4), pp. 283–292.

Shepherd, Dean A.; Patzelt, Holger; Wolfe, Marcus (2011): Moving Forward from Project Failure. Negative Emotions, Affective Commitment, and Learning from the Experience. In *Academy of Management Journal* 54 (6), pp. 1229–1259

Shih, Tse-Hua; Fan, Xitao (2008): Comparing Response Rates from Web and Mail Surveys. A Meta-Analysis. In *Field Methods* 20 (3), pp. 249–271.

Simcock, Peter; Sudbury, Lynn; Wright, Gillian (2006): Age, Perceived Risk and Satisfaction in Consumer Decision Making: A Review and Extension. In *Journal of Marketing Management* 22 (3–4), pp. 355–377.

Simonton, Dean Keith (1988): Age and Outstanding Achievement. What Do We Know After a Century of Research? In *Psychological Bulletin* 104 (2), pp. 251–267.

Sirgy, M. Joseph (1982): Self-Concept in Consumer Behavior. A Critical Review. In *Journal of Consumer Research* 9 (3), pp. 287–300.

Skinner, Jonathan (1988): Risky Income, Life Cycle Consumption, and Precautionary Savings. In *Journal of Monetary Economics* 22 (2), pp. 237–255.

Slaughter, Sarah (1993): Innovation and Learning during Implementation. A Comparison of User and Manufacturer Innovations. In *Research Policy* 22 (1), pp. 81–95.

Smith, Adam (1778): An Inquiry into the Nature and Causes of the Wealth of Nations. 2nd ed. 2 volumes. London: Strahan & Cadell.

Sorce, Patricia (1995): Cognitive Competence of Older Consumers. In *Psychology & Marketing* 12 (6), pp. 467–480.

Späth, Lothar (2008): TOP 100 2008. Die 100 innovativsten Unternehmen im Mittelstand. With assistance of Nikolaus Franke. Frankfurt am Main: Redline Wirtschaft.

Statistik Austria (2011): ISCO 08. gemeinsame deutschsprachige Titel und Erläuterungen. auf Basis der englischsprachigen Version 1.5a von April 2011. Available online at www.statistik.at/web_de/klassifikationen/oeisco08_ implementierung/informationen_zur_isco08/index.html, updated on 18/04/2011, checked on 08/10/2012.

Statistische Ämter des Bundes und der Länder: Mikrozensus 2011 und Arbeitskräftestichprobe 2011 der Europäischen Union. Stichprobenerhebung über die Bevölkerung und den Arbeitsmarkt. Fragebogen Muster. Edited by Statistische Ämter des Bundes und der Länder. Available online at https://www.destatis.de/DE/ZahlenFakten/GesellschaftStaat/Bevoelkerung/MikrozensusFragebogenMuster.pdf?__blob=publicationFile, checked on 20/02/2012.

Statistische Ämter des Bundes und der Länder (2011): Sample-based Household Survey of the 2011 Census. Edited by Statistische Ämter des Bundes und der Länder. Available online at https://www.zensus2011.de/SharedDocs/Downloads/DE/Fragebogen_International/International_Haushaltebefragung/Ausfuellhilfe_Haushaltebefragung_englisch.pdf?__blob=publicationFile&v=4, checked on 08/02/2013.

Statistisches Bundesamt Deutschland (2009): Bevölkerung Deutschlands bis 2060. 12. koordinierte Bevölkerungsvorausberechnung. Statistisches Bundesamt Deutschland. Available online at https://www.destatis.de/bevoelkerungspyramide/, checked on 16/06/2013.

Statistisches Bundesamt Deutschland (2011): Demografischer Wandel in Deutschland. Bevölkerungs- und Haushaltsentwicklung im Bund und in den Ländern. Edited by Statistische Ämter des Bundes und der Länder. Wiesbaden (Demografischer Wandel - Hintergründe und Herausforderungen, Heft 1).

Statistisches Bundesamt Deutschland (Ed.) (2012a): Bevölkerungsfortschreibung 2010. Wiesbaden (Fachserie 1, Reihe 1.3).

Statistisches Bundesamt Deutschland (2012b): Monatserhebung im Tourismus. Mai 2012. Edited by Statistisches Bundesamt Deutschland. Wiesbaden (Binnenhandel, Gastgewerbe, Tourismus).

Statistisches Bundesamt Deutschland (Ed.) (2013): Time Use Survey in Germany 2001/02. Time Use of Pensioners and the Full-time Employed. Available online at https://www.destatis.de/EN/FactsFigures/SocietyState/IncomeConsumptionLiving Conditions/TimeUse/Tables/TimeUsePensioners_ZBE.html, updated in 2013, checked on 16/03/2013.

Staudinger, Ursula M.; Häfner, Heinz (Eds.) (2008): Was ist Alter(n)? Neue Antworten auf eine scheinbar einfache Frage. Berlin: Springer (Schriften der Mathematisch-naturwissenschaftlichen Klasse der Heidelberger Akademie der Wissenschaften, 18).

Steenkamp, Jan-Benedict E. M.; Hofstede, Frenkel ter; Wedel, Michel (1999): A Cross-National Investigation into the Individual and National Cultural Antecedents of Consumer Innovativeness. In *Journal of Marketing* 63 (2), pp. 55–69.

Stephens, Nancy (1991): Cognitive Age. A Useful Concept for Advertising? In *Journal of Advertising* 20 (4), pp. 37–48.

Stone, M. (1974): Cross-Validatory Choice and Assessment of Statistical Predictions. In *Journal of the Royal Statistical Society B* 36 (1), pp. 111–147.

Sudbury, Lynn; Simcock, Peter (2009): Understanding Older Consumers through Cognitive Age and the List of Values. A U.K.-based Perspective. In *Psychology & Marketing* 26 (1), pp. 22–38.

Sudbury, Lynn; Simcock, Peter (2011): Bargain Hunting Belongers and Positive Pioneers. Key Silver Market Segments in the UK. In Florian Kohlbacher, Cornelius Herstatt (Eds.): The Silver Market Phenomenon. Marketing and Innovation in the Aging Society. 2nd ed. Berlin, Heidelberg: Springer, pp. 195–201.

Super, Donald E. (1994): Der Lebenszeit-, Lebensraumansatz der Laufbahnentwicklung. In Duane Brown, Linda Brooks (Eds.): Karriere-Entwicklung. With assistance of Maren Klostermann. Stuttgart: Klett-Cotta, pp. 211–280.

Szmigin, Isabelle; Carrigan, Marylyn (2000): The Older Consumer as Innovator. Does Cognitive Age Hold the Key? In *Journal of Marketing Management* 16 (5), pp. 505–527.

Szmigin, Isabelle; Carrigan, Marylyn (2001): Leisure and Tourism Services and the Older Innovator. In *Service Industries Journal* 21 (3), pp. 113–129.

Taylor, C. T.; Silberston, Aubrey (1973): The Economic Impact of the Patent System. A Study of the British Experience. Cambridge: Cambridge University Press.

Tempest, Sue; Barnatt, Christopher; Coupland, Christine (2008): Grey Power. Older Workers as Older Customers. In Florian Kohlbacher, Cornelius Herstatt (Eds.): The Silver Market Phenomenon. Business Opportunities in an Era of Demographic Change. 1st ed. Berlin, Heidelberg: Springer, pp. 243–255.

Tenenhaus, Michel; Amato, Silvano; Esposito Vinzi, Vincenzo (2004): A Global Goodness-of-Fit Index for PLS Structural Equation Modelling. In: Proceedings of the XLII SIS (Italian Statistical Society) Scientific Meeting, Contributed Papers. Padova, Italy: CLEUP, pp. 739–742.

Tenenhaus, Michel; Esposito Vinzi, Vincenzo; Chatelin, Yves-Marie; Lauro, Carlo (2005): PLS Path Modeling. In *Computational Statistics & Data Analysis* 48 (1), pp. 159–205.

The Camping and Caravanning Club (Ed.) (2013): Club History. Over 100 years of The Camping and Caravanning Club. Available online at www.campingandcaravanningclub.co.uk/aboutus/history/, updated in 2013, checked on 08/03/2013.

The Caravan Club Limited (Ed.) (2012): The History of The Club. Available online at www.caravanclub.co.uk/about-us/who-we-are/the-history-of-the-club/, updated in 2012, checked on 08/03/2013.

Thünker, Arnold (1999): Mit Sack und Pack und Gummiboot. Die Geschichte des Campings. Leipzig: Kiepenheuer.

Tietz, Robert; Morrison, Pamela D.; Lüthje, Christian; Herstatt, Cornelius (2005): The Process of User-Innovation. A Case Study in a Consumer Goods Setting. In *International Journal of Product Development* 2 (4), pp. 321–338.

Tinz, Teresa Valerie (2007): Spitzenprodukte durch Spitzensportler? Kooperative Produktentwicklung bei Sportartikeln. Dissertation. Universität Zürich, Zürich. Wirtschaftswissenschaftliche Fakultät.

Tongren, Hale N. (1988): Determinant Behavior Characteristics of Older Consumers. In *Journal of Consumer Affairs* 22 (1), pp. 136–157.

United Nations (1999): The World at Six Billion. Edited by United Nations. Department of Economic and Social Affairs, Population Division. Available online at www.un.org/esa/population/publications/sixbillion/sixbillion.htm, checked on 18/06/2013.

United Nations (2010): World Population Ageing, 2009. New York, NY: United Nations (Population studies (ST/ESA/SER. A), 295).

United Nations (2012): Population Ageing and Development. Ten Years after Madrid. Edited by United Nations. Department of Economic and Social Affairs, Population Division (Population Facts, 4). Available online at www.un.org/en/development/desa/population/publications/pdf/popfacts/popfacts_2012-4.pdf, checked on 16/06/2013.

United Nations (2013): World Population Prospects. The 2012 Revision. Edited by United Nations. Department of Economic and Social Affairs, Population Division. Available online at http://esa.un.org/unpd/wpp/index.htm, updated on 13/06/2013, checked on 16/06/2013.

United States Census Bureau (Ed.): American Fact Finder. Available online at http://factfinder2.census.gov/faces/nav/jsf/pages/index.xhtml, checked on 27/02/2013.

United States Census Bureau (2011): Detailed Tables on Wealth and Asset Ownership. 2011. Available online at www.census.gov/people/wealth/data/dtables.html, updated on 26/02/2013, checked on 21/06/2013.

United States Census Bureau (2013): World Population. World Population: 1950-2050. United States Census Bureau. Available online at www.census.gov/population/international/data/worldpop/graph_population.php, updated on 19/12/2013, checked on 27/01/2014.

Urban, Glen I.; Hippel, Eric von (1988): Lead User Analyses for the Development of New Industrial Products. In *Management Science* 34 (5), pp. 569–582.

Usui, Chikako (2008): Japan's Demographic Changes, Social Implications, and Business Opportunities. In Florian Kohlbacher, Cornelius Herstatt (Eds.): The Silver Market Phenomenon. Business Opportunities in an Era of Demographic Change. 1st ed. Berlin, Heidelberg: Springer, pp. 71–82.

Usui, Chikako (2011): Japan's Population Aging and Silver Industries. In Florian Kohlbacher, Cornelius Herstatt (Eds.): The Silver Market Phenomenon. Marketing and Innovation in the Aging Society. 2nd ed. Berlin, Heidelberg: Springer, pp. 325–337.

Vanderwerf, Pieter A. (1990): Product Tying and Innovation in U.S. Wire Preparation Equipment. In *Research Policy* 19 (1), pp. 83–96.

Venkatraman, Meera P. (1991): The Impact of Innovativeness and Innovation Type on Adoption. In *Journal of Retailing* 67 (1), p. 51.

Verworn, Birgit; Schwarz, Doreen; Herstatt, Cornelius (2009): Changing Workforce Demographics. Strategies Derived from the Resource-Based View of HRM. In *International Journal of Human Resources Development and Management* 9 (2), pp. 149–161.

Voss, Christopher A. (1985): The Role of Users in the Development of Applications Software. In *Journal of Product Innovation Management* 2 (2), pp. 113–121.

Weiber, Rolf; Mühlhaus, Daniel (2010): Strukturgleichungsmodellierung. Eine anwendungsorientierte Einführung in die Kausalanalyse mit Hilfe von AMOS, SmartPLS und SPSS. Berlin, Heidelberg: Springer.

Westfalia Mobil GmbH (Ed.) (2013): Westfalia - Company History. Mobile Passion with History. Available online at www.westfalia-mobil.net/en/unternehmen/unternehmenshistorie.php, updated in 2013, checked on 12/03/2013.

Wikipedia contributors (2013a): Thomas Edison. Edited by The Free Encyclopedia Wikipedia. Available online at http://en.wikipedia.org/w/index.php?title=Thomas_Edison&oldid=561800994, updated on 27/06/2013, checked on 29/06/2013.

Wikipedia contributors (2013b): Camping. Edited by The Free Encyclopedia Wikipedia. Available online at http://en.wikipedia.org/w/index.php?title=Camping&oldid= 565614947, updated on 24/07/2013, checked on 31/07/2013.

Wikipedia contributors (2014): Best Agers. Edited by The Free Encyclopedia Wikipedia. Available online at http://de.wikipedia.org/w/index.php?title=Best_Ager&oldid= 126185274, updated on 06/01/2014, checked on 28/01/2014.

Wilkes, Robert E. (1992): A Structural Modeling Approach to the Measurement and Meaning of Cognitive Age. In *Journal of Consumer Research* 19 (2), pp. 292–301.

Williams, Larry J.; Edwards, Jeffrey R.; Vandenberg, Robert J. (2003): Recent Advances in Causal Modeling Methods for Organizational and Management Research. In *Journal of Management* 29 (6), pp. 903–936.

Wilson, Robert Woodrow (1975): The Sale of Technology through Licensing. Dissertation. Yale University.

Wold, Herman Ole Andreas (1982): Soft Modelling. The Basic Design and Some Extensions. In K. G. Jöreskog, Herman Ole Andreas Wold (Eds.): Systems under Indirect Observation. Causality, Structure, Prediction. Part 2. Amsterdam, New York, Oxford: North-Holland (Contributions to Economic Analysis, 139), pp. 1–54.

Wolfe, David B. (1994): Targeting the Mature Mind. In *American Demographics* 16 (3), pp. 32–36.

World Health Organization (2013): Global Health Observatory Data Repository. Life Expectancy by Country. Edited by World Health Organization. Available online at http://apps.who.int/gho/data/node.main.688?lang=en, updated in 2013, checked on 10/12/2013.

World Health Organization & US National Institute of Aging (Ed.) (2011): Global Health and Aging. Available online at www.who.int/ageing/publications/global_health/ en/index.html, checked on 28/06/2013.

World Intellectual Property Organization (Ed.): WIPO Intellectual Property Handbook. Policy, Law and Use. 2nd ed. Geneva.

Zaichkowsky, Judith Lynne (1985): Measuring the Involvement Construct. In *Journal of Consumer Research* 12 (3), pp. 341–352.

Appendix

Appendix 1   List of Interviewees and Experts

Name	Institution	Area of Expertise
Initial interview partners		
Daniela Leipelt	BVCD e.V.	German camping market
Norbert Gröll	Klappcaravanforum.de	Camping community administrator
Gerlinde Jaensch	Naturpark-Camping Prinzenholz	Campsite manager
Mr. Lemke	SolarMaxiPower	User manufacturer
Experts to Evaluate Impact of Age		
Dr. Katrin Claßen	Heidelberg University Institute of Psychology	Psychological aging, technology acceptance
Prof. Dr. Josefine Heusinger	Institut für Gerontologische Forschung e.V.	
Dipl.-Des. Dipl.-Ing. Mathias Knigge	grauwert	Product design for aging users
Dr. Florian Kohlbacher	German Institute for Japanese Studies	Silver Market
Dr. Tim Schweisfurth	Technical University of Munich School of Management	User innovation, lead user
Prof. Dr. Clemens Tesch-Römer	German Centre of Gerontology (Deutsches Zentrum für Altersfragen)	Comparative aging research

Appendix 2 Tests for Mode Effects of Data Collection Method on Measurement

Table 35: Results of Test for Mode Effects on Measurement

Item	Matching Sample 1 Age = 43…50; Income = 2500…3500; Education = 2…6						Matching Sample 2 Age = 51…58; Income = 2500…3500; Education = 2…6					
	Mann-Whitney- U	Wilcoxon- W	Z	Asymp. Sig. (2-tail'd)	Kolmo-gorov-Smirnov -Z	Asymp. Sig. (2-tail'd)	Mann-Whitney- U	Wilcoxon- W	Z	Asymp. Sig. (2-tail'd)	Kolmo-gorov-Smirnov -Z	Asymp. Sig. (2-tail'd)
LU [1]	180.5	225.5	-0.428	0.669	0.393	0.998	79.5	199.5	-0.541	0.588	0.602	0.861
LU [2]	194.5	239.5	-0.088	0.930	0.207	1.000	70.5	148.5	-1.029	0.303	0.689	0.730
LU [3]	189.0	1179.0	-0.465	0.642	0.242	1.000	89.0	209.0	-0.107	0.914	0.172	1.000
LU [4]	147.5	192.5	-1.422	0.155	0.752	0.623	82.5	202.5	-0.504	0.614	0.430	0.993
LU [5]	130.5	175.5	-1.651	0.099	0.932	0.350	58.0	178.0	-1.666	0.096	0.818	0.516
LU [6]	165.5	210.5	-0.789	0.430	0.518	0.951	62.0	182.0	-1.437	0.151	0.861	0.449
Motivation [1]	12.0	13.0	-0.283	0.777	0.849	0.467	17.5	32.5	-0.791	0.429	0.219	1.000
Motivation [2]	12.5	13.5	-0.196	0.845	0.877	0.426	20.0	35.0	0.000	1.000	0.000	1.000
Motivation [3]	12.5	363.5	-0.196	0.845	0.387	0.998	20.0	35.0	0.000	1.000	0.000	1.000
Motivation [4]	2.5	3.5	-1.408	0.159	1.084	0.191	18.5	54.5	-0.233	0.816	0.395	0.998
Motivation [5]	10.0	361.0	-0.531	0.596	0.849	0.467	12.0	27.0	-1.868	0.062	0.702	0.708
PK [1]	154.0	199.0	-1.098	0.272	0.849	0.467	81.0	201.0	-0.469	0.639	0.258	1.000
PK [2]	130.5	175.5	-1.679	0.093	0.849	0.467	57.0	177.0	-1.679	0.093	0.689	0.730
PK [3]	159.0	1149.0	-0.976	0.329	1.332	0.057	72.5	192.5	-0.920	0.357	0.473	0.978
TE [1]	117.5	162.5	-2.038	0.042	0.511	0.957	38.5	158.5	-2.707	0.007	1.377	0.045
TE [2]	133.5	178.5	-1.610	0.107	1.311	0.064	61.5	181.5	-1.479	0.139	0.947	0.332
TE [3]	176.5	221.5	-0.543	0.587	1.311	0.064	35.0	155.0	-3.037	0.002	1.506	0.021
TE [4]	135.5	180.5	-1.553	0.120	0.697	0.716	34.5	154.5	-2.860	0.004	1.377	0.045
UE [1]	142.5	187.5	-1.331	0.183	0.683	0.739	71.5	191.5	-0.911	0.362	0.861	0.449
UE [2]	188.0	233.0	-0.237	0.812	0.414	0.995	75.5	153.5	-0.711	0.477	0.775	0.586
IB [1]	103.0	148.0	-2.599	0.009	0.538	0.934	60.0	180.0	-1.690	0.091	0.861	0.449
IB [2]	115.0	160.0	-2.142	0.032	0.456	0.986	53.0	173.0	-1.982	0.047	0.947	0.332
Chronological Age	136.0	1126.0	-1.494	0.135	0.607	0.854	58.0	136.0	-1.586	0.113	1.248	0.089
FEEL Age	118.0	1108.0	-1.975	0.048	0.518	0.951	42.5	120.5	-2.389	0.017	1.291	0.071
LOOK Age	172.5	1162.5	-0.647	0.518	0.521	0.949	85.5	163.5	-0.237	0.813	0.215	1.000
DO Age	147.5	1137.5	-1.239	0.215	0.891	0.406	65.0	143.0	-1.255	0.210	1.119	0.163
INTEREST Age	157.5	1147.5	-1.010	0.313	0.179	1.000	81.5	159.5	-0.427	0.669	0.430	0.993
Evaluation Age	184.0	1174.0	-0.357	0.721	0.226	1.000	84.5	162.5	-0.289	0.773	0.645	0.799
Income [1]	160.5	1150.5	-1.026	0.305	0.304	1.000	66.0	186.0	-1.352	0.176	0.689	0.730
Income [2]	131.5	176.5	-1.081	0.280	0.289	1.000	88.5	166.5	-0.074	0.941	0.516	0.952
Time	131.5	176.5	-1.368	0.171	0.612	0.849	79.5	199.5	-0.517	0.605	0.430	0.993
Education	196.5	241.5	-0.038	0.969	0.393	0.998	62.5	140.5	-1.437	0.151	0.818	0.516
Job	178.5	223.5	-0.580	0.562	0.207	1.000	54.0	174.0	-2.420	0.016	1.033	0.236
Family	170.0	1160.0	-1.018	0.309	0.242	1.000	75.0	153.0	-1.611	0.107	0.430	0.993
GDR	173.0	218.0	-0.839	0.402	0.752	0.623	72.0	192.0	-1.612	0.107	0.516	0.952
Gender	131.5	167.5	-1.588	0.112	0.932	0.350	61.5	181.5	-1.831	0.067	0.818	0.516

Grouping Variable: Survey type

Appendix 3 Testing for Measurement Invariance – Chronological Age Groups

Table 36: Indicator and Construct Reliability

		Silver Age Group				Non-Silver Age Group			
		Outer Loading	T-Value	Composite Reliability	AVE	Outer Loading	T-Value	Composite Reliability	AVE
Critical Value		$\lambda \geq 0.7$	≥1.65 : p<0.10 ≥1.96 : p<0.05 ≥2.58 : p<0.01	CR ≥ 0.7	AVE ≥ 0.6	$\lambda \geq 0.7$	≥1.65 : p<0.10 ≥1.96 : p<0.05 ≥2.58 : p<0.01	CR ≥ 0.7	AVE ≥ 0.6
Construct	Item								
Use Experience[†]	UE [1]	-0.016	0.035	n/a	n/a	0.516	2.879	n/a	n/a
	UE [2]	0.998	2.599			0.930	11.990		
Product Knowledge	PK [1]	0.776	12.314	0.866	0.683	0.659	10.093	0.813	0.594
	PK [2]	0.834	15.698			0.829	28.709		
	PK [3]	0.868	23.629			0.812	22.893		
Technical Expertise	TE [1]	0.922	46.427	0.951	0.828	0.895	38.156	0.945	0.812
	TE [2]	0.900	36.142			0.876	50.770		
	TE [3]	0.880	28.309			0.903	55.978		
	TE [4]	0.936	81.517			0.930	89.857		
Ahead of Trend	LU [1]	0.811	16.258	0.854	0.661	0.776	11.549	0.837	0.632
	LU [2]	0.860	19.297			0.764	11.293		
	LU [4]	0.767	14.678			0.844	19.120		
Exp. High Benefits	LU [5]	0.885	19.938	0.871	0.771	0.928	71.681	0.885	0.793
	LU [6]	0.871	15.602			0.852	24.798		
Innovative Behavior[†]	IB [1]	1.000	n/a	n/a	n/a	1.000	n/a	n/a	n/a

† Formative construct   Cases: 110; Samples: 5,000   Cases: 223; Samples: 5,000

Table 37: PLS Cross-Loadings

	Silver Age Group						Non-Silver Age Group					
Item	UE[†]	PK	TE	AoT	EHB	IB[†]	UE[†]	PK	TE	AoT	EHB	IB[†]
UE [1]	-0.016	0.021	-0.108	-0.042	-0.114	0.034	**0.516**	0.225	-0.062	0.095	-0.042	0.135
UE [2]	**0.998**	0.226	0.234	0.152	0.028	0.130	**0.930**	0.245	0.155	0.092	0.127	0.350
PK [1]	0.125	**0.776**	0.245	0.230	0.017	0.228	0.248	**0.659**	0.233	0.144	0.066	0.155
PK [2]	0.233	**0.834**	0.259	0.303	0.049	0.110	0.237	**0.829**	0.317	0.396	0.267	0.244
PK [3]	0.196	**0.868**	0.564	0.303	0.058	0.217	0.219	**0.812**	0.453	0.238	0.217	0.266
TE [1]	0.269	0.461	**0.922**	0.240	0.213	0.366	0.054	0.373	**0.895**	0.193	0.171	0.294
TE [2]	0.228	0.464	**0.900**	0.373	0.192	0.385	0.173	0.458	**0.876**	0.281	0.173	0.290
TE [3]	0.125	0.391	**0.880**	0.194	0.134	0.282	0.059	0.385	**0.903**	0.226	0.169	0.228
TE [4]	0.238	0.389	**0.936**	0.340	0.245	0.429	0.103	0.386	**0.930**	0.254	0.221	0.339
LU [1]	0.146	0.225	0.251	**0.811**	0.367	0.102	-0.028	0.217	0.155	**0.776**	0.216	-0.007
LU [2]	0.130	0.330	0.253	**0.860**	0.236	0.229	0.031	0.248	0.197	**0.764**	0.133	0.001
LU [4]	0.106	0.273	0.279	**0.767**	0.376	0.236	0.197	0.347	0.260	**0.844**	0.299	0.207
LU [5]	0.031	0.093	0.162	0.395	**0.885**	0.256	0.116	0.273	0.225	0.312	**0.928**	0.462
LU [6]	0.032	-0.002	0.225	0.316	**0.871**	0.236	0.041	0.173	0.125	0.187	**0.852**	0.360
IB [1]	0.128	0.227	0.408	0.236	0.280	**1.000**	0.354	0.297	0.321	0.119	0.468	**1.000**

† Formative construct

Table 38: Outer Loadings, Weights, and Multicollinearity of Formative Constructs

		Outer Loading	T-Value	Outer Weight	T-Value	VIF	Correlation[†]
Critical Value		$\lambda \geq 0.5$	≥1.65 : p<0.10 ≥1.96 : p<0.05 ≥2.58 : p<0.01	$\lambda \geq 0.5$	≥1.65 : p<0.10 ≥1.96 : p<0.05 ≥2.58 : p<0.01	VIF < 5	
Group	Item						
Silver Age	UE [1]	-0.016	0.035	-0.069	0.145	1.003	0.053[n.s.]
	UE [2]	0.998	2.599	1.001	2.471	1.003	
Non-Silver Age	UE [1]	0.516	2.879	0.372	2.014	1.028	0.165**
	UE [2]	0.930	11.990	0.869	8.014	1.028	

† Pearson correlation coefficient
** Pearson correlation significant with p < 0.05 (2-tailed)

Appendix 4   Testing for Measurement Invariance – Cognitive Age Groups

Table 39:   Indicator and Construct Reliability

		High Cognitive Age Group				Low Cognitive Age Group			
		Outer Loading	T-Value	Composite Reliability	AVE	Outer Loading	T-Value	Composite Reliability	AVE
Critical Value		$\lambda \geq 0.7$	≥1.65 : p<0.10 ≥1.96 : p<0.05 ≥2.58 : p<0.01	$CR \geq 0.7$	$AVE \geq 0.6$	$\lambda \geq 0.7$	≥1.65 : p<0.10 ≥1.96 : p<0.05 ≥2.58 : p<0.01	$CR \geq 0.7$	$AVE \geq 0.6$
Construct	Item								
Use Experience†	UE [1] UE [2]	0.101 1.000	0.308 4.271	n/a	n/a	0.483 0.917	1.178 3.232	n/a	n/a
Product Knowledge	PK [1] PK [2] PK [3]	0.734 0.830 0.838	11.383 19.834 20.067	0.844	0.643	0.663 0.841 0.804	6.663 21.791 16.565	0.815	0.598
Technical Expertise	TE [1] TE [2] TE [3] TE [4]	0.912 0.879 0.879 0.936	44.294 32.924 31.024 89.768	0.946	0.813	0.866 0.851 0.875 0.917	20.102 33.507 28.147 47.550	0.931	0.770
Ahead of Trend	LU [1] LU [2] LU [4]	0.812 0.862 0.779	16.336 23.010 16.362	0.859	0.670	0.799 0.812 0.813	9.882 9.810 13.209	0.849	0.653
Exp. High Benefits	LU [5] LU [6]	0.922 0.864	39.555 18.649	0.888	0.798	0.935 0.850	56.739 17.836	0.888	0.799
Innovative Behavior†	IB [1]	1.000	n/a	n/a	n/a	1.000	n/a	n/a	n/a
† Formative construct		Cases: 128; Samples: 5,000				Cases: 119; Samples: 5,000			

Table 40:   PLS Cross-Loadings

	High Cognitive Age Group						Low Cognitive Age Group					
Item	UE†	PK	TE	AoT	EHB	IB†	UE†	PK	TE	AoT	EHB	IB†
UE [1]	0.101	0.055	-0.078	0.008	-0.157	0.028	**0.483**	0.225	-0.003	0.077	-0.107	0.045
UE [2]	**1.000**	0.312	0.304	0.213	0.041	0.181	**0.917**	0.222	0.176	0.135	0.154	0.216
PK [1]	0.189	**0.734**	0.248	0.222	-0.030	0.184	0.258	**0.663**	0.199	0.133	0.046	0.152
PK [2]	0.279	**0.830**	0.302	0.303	0.103	0.158	0.226	**0.841**	0.358	0.395	0.215	0.287
PK [3]	0.272	**0.838**	0.486	0.277	0.069	0.165	0.208	**0.804**	0.488	0.205	0.197	0.338
TE [1]	0.323	0.425	**0.913**	0.214	0.197	0.355	0.060	0.417	**0.866**	0.224	0.248	0.345
TE [2]	0.289	0.421	**0.879**	0.357	0.227	0.319	0.210	0.509	**0.851**	0.332	0.256	0.374
TE [3]	0.190	0.392	**0.879**	0.200	0.163	0.297	0.083	0.311	**0.875**	0.194	0.255	0.308
TE [4]	0.281	0.378	**0.936**	0.328	0.262	0.423	0.162	0.411	**0.917**	0.253	0.346	0.434
LU [1]	0.175	0.225	0.251	**0.812**	0.302	0.116	0.010	0.211	0.187	**0.799**	0.255	0.064
LU [2]	0.197	0.307	0.220	**0.862**	0.176	0.197	0.079	0.301	0.216	**0.812**	0.165	0.158
LU [4]	0.153	0.287	0.279	**0.779**	0.338	0.235	0.227	0.291	0.286	**0.813**	0.252	0.231
LU [5]	-0.001	0.092	0.214	0.350	**0.922**	0.308	0.113	0.217	0.356	0.269	**0.935**	0.556
LU [6]	0.083	0.020	0.214	0.248	**0.864**	0.222	0.040	0.169	0.182	0.225	**0.850**	0.384
IB [1]	0.181	0.208	0.390	0.229	0.302	**1.000**	0.208	0.354	0.422	0.200	0.541	**1.000**

† Formative construct

Table 41:   Outer Loadings, Weights, and Multicollinearity of Formative Constructs

		Outer Loading	T-Value	Outer Weight	T-Value	VIF	Correlation†
Critical Value		$\lambda \geq 0.5$	≥1.65 : p<0.10 ≥1.96 : p<0.05 ≥2.58 : p<0.01	$\lambda \geq 0.5$	≥1.65 : p<0.10 ≥1.96 : p<0.05 ≥2.58 : p<0.01	VIF < 5	
Group	Item						
High Cognitive Age	UE [1] UE [2]	0.101 1.000	0.308 4.271	0.008 0.999	0.022 3.957	1.009 1.009	0.093[n.s.]
Low Cognitive Age	UE [1] UE [2]	0.483 0.917	1.178 3.232	0.401 0.880	0.940 2.748	1.009 1.009	0.093[n.s.]

† Pearson correlation coefficient

Appendix 5 Results of PLS-MGA with Sub-Dimensions of Cognitive Age

Table 42: PLS-MGA for FEEL Age and LOOK Age Groups

Exogenous Variable	Endogenous Variable	High FEEL Age Group γ^{FEEL+}	Low FEEL Age Group γ^{FEEL-}	Group Differences Δ of γ	p-value	High LOOK Age Group γ^{LOOK+}	Low LOOK Age Group γ^{LOOK-}	Group Differences Δ of γ	p-value
Use Experience	PK	0.220***	0.219***	0.001	0.511	0.147$^{n.s.}$	0.251***	-0.104	0.233
	AoT	0.060$^{n.s.}$	-0.020$^{n.s.}$	0.080	0.719	0.096$^{n.s.}$	-0.010$^{n.s.}$	0.106	0.760
	HEB	-0.037$^{n.s.}$	0.059$^{n.s.}$	-0.096	0.255	-0.010$^{n.s.}$	-0.008$^{n.s.}$	-0.002	0.492
	IB	0.125$^{n.s.}$	0.170***	-0.045	0.343	-0.016$^{n.s.}$	0.258***	-0.275	0.007
Product Knowledge	AoT	0.204***	0.287***	-0.084	0.231	0.236***	0.312***	-0.076	0.244
	HEB	-0.068$^{n.s.}$	0.066$^{n.s.}$	-0.134	0.145	-0.068$^{n.s.}$	0.087$^{n.s.}$	-0.155	0.107
	IB	0.040$^{n.s.}$	0.033$^{n.s.}$	0.008	0.521	0.034$^{n.s.}$	0.073$^{n.s.}$	-0.039	0.356
Technical Expertise	PK	0.364***	0.477***	-0.113	0.150	0.419***	0.434***	-0.014	0.451
	AoT	0.225***	0.117$^{n.s.}$	0.108	0.849	0.182**	0.103$^{n.s.}$	0.079	0.774
	HEB	0.145*	0.162*	-0.017	0.448	0.157**	0.085$^{n.s.}$	0.072	0.696
	IB	0.283***	0.289***	-0.006	0.480	0.348***	0.200***	0.147	0.921
Ahead of Trend	HEB	0.295***	0.248***	0.048	0.646	0.345***	0.234***	0.112	0.815
	IB	0.000$^{n.s.}$	-0.100$^{n.s.}$	0.100	0.792	0.021$^{n.s.}$	-0.133*	0.153	0.898
High Exp. Benefits	IB	0.235***	0.399***	-0.164	0.075	0.245***	0.440***	-0.195	0.039

* p < 0.10; ** p < 0.05; *** p < 0.01
Cases: 146   Cases: 147
Samples: 5,000
Median: 42 y

Cases: 122   Cases: 173
Samples: 5,000
Median: 47 y

Table 43: PLS-MGA for DO Age and INTEREST Age Groups

Exogenous Variable	Endogenous Variable	High DO Age Group γ^{DO+}	Low DO Age Group γ^{DO-}	Group Differences Δ of γ	p-value	High INTEREST Age Group $\gamma^{INTEREST+}$	Low INTEREST Age Group $\gamma^{INTEREST-}$	Group Differences Δ of γ	p-value
Use Experience	PK	0.193**	0.216***	-0.024	0.434	0.216***	0.251***	-0.035	0.400
	AoT	0.069$^{n.s.}$	0.015$^{n.s.}$	0.055	0.645	0.039$^{n.s.}$	0.084$^{n.s.}$	-0.045	0.375
	HEB	-0.019$^{n.s.}$	0.064$^{n.s.}$	-0.084	0.280	-0.081$^{n.s.}$	0.092$^{n.s.}$	-0.173	0.130
	IB	0.112$^{n.s.}$	0.157*	-0.045	0.352	0.083$^{n.s.}$	0.138*	-0.055	0.323
Product Knowledge	AoT	0.174**	0.222***	-0.048	0.338	0.281***	0.204***	0.077	0.769
	HEB	-0.098$^{n.s.}$	0.068$^{n.s.}$	-0.167	0.090	-0.086$^{n.s.}$	0.097$^{n.s.}$	-0.183	0.061
	IB	0.098$^{n.s.}$	0.119$^{n.s.}$	-0.021	0.425	0.011$^{n.s.}$	0.146*	-0.135	0.115
Technical Expertise	PK	0.397***	0.386***	0.011	0.543	0.399***	0.387***	0.012	0.548
	AoT	0.226***	0.129*	0.097	0.811	0.168**	0.185**	-0.018	0.430
	HEB	0.156*	0.151*	0.005	0.517	0.194**	0.131$^{n.s.}$	0.064	0.698
	IB	0.200***	0.281***	-0.082	0.245	0.309***	0.289***	0.019	0.572
Ahead of Trend	HEB	0.397***	0.252***	0.146	0.884	0.306***	0.310***	-0.003	0.491
	IB	-0.007$^{n.s.}$	-0.101$^{n.s.}$	0.094	0.788	-0.032$^{n.s.}$	-0.057$^{n.s.}$	0.025	0.584
High Exp. Benefits	IB	0.297***	0.337***	-0.040	0.368	0.274***	0.366***	-0.092	0.206

* p < 0.10; ** p < 0.05; *** p < 0.01
Cases: 131   Cases: 152
Samples: 5,000
Median: 42 y

Cases: 142   Cases: 137
Samples: 5,000
Median: 42 y

Table 44: PLS-MGA for Age Difference Groups in the Full and the Silver Market Sample

		Full Sample				Silver Market Sample			
		High Age Difference Group $\gamma_{AgeDiff+}$	Low Age Difference Group $\gamma_{AgeDiff-}$	Group Differences		High Age Difference Group $\gamma_{AgeDiff+}$	Low Age Difference Group $\gamma_{AgeDiff-}$	Group Differences	
Exogenous Variable	Endogenous Variable			Δ of γ	p-value			Δ of γ	p-value
Use Experience	PK	0.285**	0.174*	0.111	0.761	0.006$^{n.s.}$	0.003$^{n.s.}$	0.002	0.532
	AoT	0.038$^{n.s.}$	-0.049$^{n.s.}$	0.087	0.730	0.007$^{n.s.}$	0.002$^{n.s.}$	0.005	0.551
	HEB	-0.031$^{n.s.}$	0.020$^{n.s.}$	-0.051	0.383	0.001$^{n.s.}$	0.005$^{n.s.}$	-0.005	0.496
	IB	0.163$^{n.s.}$	0.212**	-0.049	0.370	0.000$^{n.s.}$	0.007$^{n.s.}$	-0.007	0.475
Product Knowledge	AoT	0.197*	0.302***	-0.105	0.174	0.135$^{n.s.}$	0.466**	-0.332	0.039
	HEB	0.022$^{n.s.}$	0.058$^{n.s.}$	-0.036	0.379	-0.694*	-0.206$^{n.s.}$	-0.488	0.011
	IB	-0.002$^{n.s.}$	0.110$^{n.s.}$	-0.112	0.159	0.014$^{n.s.}$	0.108$^{n.s.}$	-0.094	0.316
Technical Expertise	PK	0.350***	0.491***	-0.141	0.094	0.326*	0.581***	-0.255	0.099
	AoT	0.157*	0.181*	-0.023	0.406	0.202$^{n.s.}$	0.416*	-0.214	0.148
	HEB	0.170**	0.058$^{n.s.}$	0.112	0.824	0.608**	0.358$^{n.s.}$	0.250	0.892
	IB	0.285***	0.240***	0.046	0.668	0.384**	0.417*	-0.033	0.430
Ahead of Trend	HEB	0.233**	0.347***	-0.114	0.162	0.663***	0.831***	-0.168	0.202
	IB	-0.064$^{n.s.}$	-0.074$^{n.s.}$	0.010	0.539	-0.040$^{n.s.}$	0.110$^{n.s.}$	-0.150	0.253
High Exp. Benefits	IB	0.333***	0.377***	-0.045	0.339	0.133**	0.023$^{n.s.}$	0.110	0.694

* p < 0.10
** p < 0.05
*** p < 0.01

Cases: 160 Cases: 165
Samples: 5,000
Median: -6.0 y

Cases: 55 Cases: 55
Samples: 5,000
Median: -8.2 y

Appendix 6   Impact of Control Variables

Table 45: Results of Main Effects Model with Control Variables

	Exogenous Variable	Endogenous Variable	Path Coefficient	T-Value
		Critical Value	$\gamma > 0.2$	$\geq 1.65 : p < 0.10$ $\geq 1.96 : p < 0.05$ $\geq 2.58 : p < 0.01$
Main Effects	Use Experience	Ahead of Trend	0.014	0.185
		High Expected Benefits	-0.072	0.898
		Innovative Behavior	0.154	2.398
		Product Knowledge	0.334	5.158
	Product Knowledge	Ahead of Trend	0.203	3.128
		High Expected Benefits	0.091	1.348
		Innovative Behavior	0.103	1.679
	Technical Expertise	Ahead of Trend	0.212	3.000
		High Expected Benefits	0.120	1.785
		Innovative Behavior	0.263	4.210
		Product Knowledge	0.442	7.680
	Ahead of Trend	High Expected Benefits	0.234	3.439
		Innovative Behavior	-0.077	1.238
	High Expected Benefits	Innovative Behavior	0.389	6.603
Control Variables	Available Time	Ahead of Trend	0.052	0.700
		High Expected Benefits	0.049	0.616
		Innovative Behavior	0.020	0.295
		Product Knowledge	0.220	3.210
	Disposable Income	Ahead of Trend	0.045	0.621
		High Expected Benefits	0.090	1.179
		Innovative Behavior	-0.049	0.801
		Product Knowledge	0.005	0.070
	Education	Ahead of Trend	-0.015	0.229
		High Expected Benefits	0.148	2.272
		Innovative Behavior	0.083	1.386
		Product Knowledge	-0.108	1.938
	Gender	Ahead of Trend	0.025	0.373
		High Expected Benefits	0.034	0.547
		Innovative Behavior	-0.026	0.481
		Product Knowledge	-0.107	1.964
	Income	Ahead of Trend	0.184	2.869
		High Expected Benefits	-0.053	0.737
		Innovative Behavior	-0.036	0.618
		Product Knowledge	0.137	2.442
	Marital status	Ahead of Trend	0.018	0.337
		High Expected Benefits	0.162	2.574
		Innovative Behavior	-0.025	0.426
		Product Knowledge	0.036	0.826
	Occupation intensity	Ahead of Trend	0.098	1.296
		High Expected Benefits	0.128	1.685
		Innovative Behavior	0.120	1.712
		Product Knowledge	0.091	1.244
	Origin	Ahead of Trend	-0.023	0.395
		High Expected Benefits	0.028	0.533
		Innovative Behavior	0.097	1.785
		Product Knowledge	-0.007	0.104

Cases: 256
Samples: 5,000

Appendix 7 Process Characteristics of User Innovators

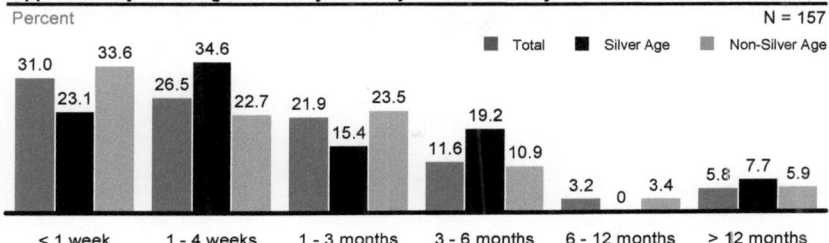

Figure 33: Development Time of User Innovators[739]

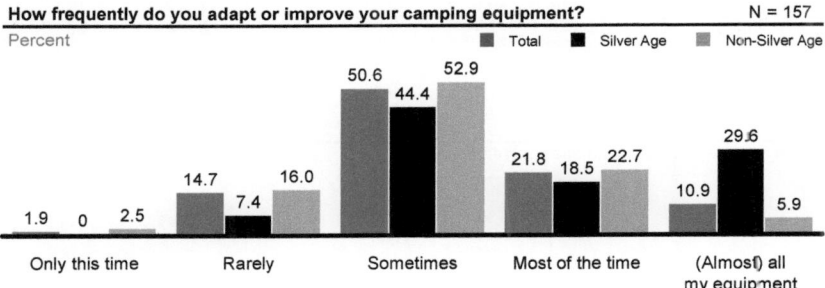

Figure 34: Development Frequency of User Innovators[740]

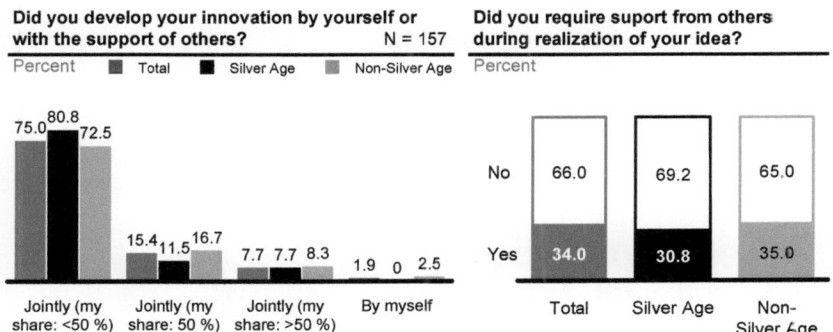

Figure 35: Cooperation during Ideation and Realization Phase[741]

---

[739] Own illustration.
[740] Own illustration.

Appendix 8 Correlation of Age Measurement with Innovation Characteristics

Table 46: Correlation Coefficients of Age Measurements with Innovation Characteristics

Innovation Characteristic	Chronol. Age	Cognitive Age	FEEL Age	LOOK Age	DO Age	INTEREST Age
PROCESS QUALITY						
Development Stage	$-0.020^{n.s.}$	$-0.063^{n.s.}$	-0.091^{*}	$-0.041^{n.s.}$	-0.099^{*}	$-0.037^{n.s.}$
Development Time	$0.006^{n.s.}$	$0.013^{n.s.}$	$-0.020^{n.s.}$	$0.015^{n.s.}$	$0.030^{n.s.}$	$0.031^{n.s.}$
Development Frequency	$0.122^{n.s.}$	$-0.025^{n.s.}$	$-0.023^{n.s.}$	$0.066^{n.s.}$	$-0.016^{n.s.}$	$-0.110^{n.s.}$
Cooperation (Ideation)	$-0.044^{n.s.}$	$-0.050^{n.s.}$	$-0.066^{n.s.}$	$-0.079^{n.s.}$	$-0.052^{n.s.}$	$0.001^{n.s.}$
Cooperation (Realization)	$-0.084^{n.s.}$	$-0.078^{n.s.}$	$-0.079^{n.s.}$	$-0.085^{n.s.}$	$-0.030^{n.s.}$	$-0.064^{n.s.}$
INNOVATION TYPE						
Comfort Improvement	$0.054^{n.s.}$	$0.034^{n.s.}$	$0.013^{n.s.}$	$0.022^{n.s.}$	$0.008^{n.s.}$	$0.052^{n.s.}$
Cost Reduction	$0.010^{n.s.}$	$-0.092^{n.s.}$	$-0.100^{n.s.}$	$0.009^{n.s.}$	-0.136^{*}	$-0.093^{n.s.}$
New Functionality	-0.197^{**}	-0.182^{**}	-0.185^{**}	-0.134^{*}	-0.142^{*}	-0.156^{*}
Time Savings	$-0.082^{n.s.}$	-0.140^{*}	-0.155^{*}	$-0.112^{n.s.}$	$-0.116^{n.s.}$	$-0.119^{n.s.}$
Improved Compatibility	$-0.69^{n.s.}$	$-0.036^{n.s.}$	$-0.013^{n.s.}$	$-0.114^{n.s.}$	$-0.018^{n.s.}$	$-0.046^{n.s.}$
INNOVATION QUALITY						
Newness	$-0.004^{n.s.}$	$-0.071^{n.s.}$	$-0.019^{n.s.}$	$-0.066^{n.s.}$	$-0.012^{n.s.}$	$-0.102^{n.s.}$
Technical Quality	$0.013^{n.s.}$	$0.021^{n.s.}$	$0.084^{n.s.}$	$0.013^{n.s.}$	$0.042^{n.s.}$	$-0.038^{n.s.}$
Creativity	$-0.058^{n.s.}$	-0.169^{**}	-0.147^{*}	$-0.128^{n.s.}$	-0.142^{*}	-0.195^{**}
Benefits for Others (today)	$-0.099^{n.s.}$	-0.133^{*}	$-0.128^{n.s.}$	$-0.103^{n.s.}$	$-0.127^{n.s.}$	$-0.105^{n.s.}$
Benefits for Others (future)	$-0.053^{n.s.}$	$-0.100^{n.s.}$	$-0.110^{n.s.}$	$-0.103^{n.s.}$	$-0.089^{n.s.}$	$-0.080^{n.s.}$
Sales Potential (today)	$0.102^{n.s.}$	$0.023^{n.s.}$	$0.003^{n.s.}$	$0.090^{n.s.}$	$0.003^{n.s.}$	$-0.025^{n.s.}$
Sales Potential (future)	$0.077^{n.s.}$	$-0.014^{n.s.}$	$-0.004^{n.s.}$	$0.033^{n.s.}$	$-0.028^{n.s.}$	$-0.059^{n.s.}$

Correlation coefficients according to Spearman's rho

* $p < 0.10$
** $p < 0.05$
*** $p < 0.01$

[741] Own illustration.

Appendix 9 Mean Comparisons of Age Groups

Appendix 10 Survey Questionnaire

Antwortbogen zum Forschungsprojekt

„Nutzerinnovatoren in Silver Markets am Beispiel Camping/Caravaning"

der TU Hamburg-Harburg,
Institut für Technologie- und Innovationsmanagement

Vielen Dank für Ihre Bereitschaft an meiner Befragung teilzunehmen, die ich im Rahmen meiner Promotion an der Technischen Universität Hamburg-Harburg durchführe. Durch Ihre Mithilfe unterstützen Sie nicht nur meine Forschung, sondern durch die gewonnenen Erkenntnisse auch Unternehmen dabei bessere und vor allem bedarfsgerechtere Campingprodukte zu entwickeln.

Ich untersuche, inwiefern Camping- und Caravaning-Touristen ihre Fahrzeuge und Ausrüstungsgegenstände modifizieren oder sogar komplett selbst entwickeln und welche Faktoren dieses Verhalten beeinflussen.

Ich versichere Ihnen, dass alle von Ihnen gemachten Angaben streng vertraulich und anonym behandelt und ausschließlich zum Zwecke meiner Dissertation verwendet werden. Es handelt sich um ein wissenschaftliches Projekt ohne kommerzielles Interesse. Sollten Sie Fragen und Anmerkungen haben, erreichen Sie mich unter konstantin.wellner@tuhh.de.

Die Beantwortung aller Fragen des Fragebogens dauert erfahrungsgemäß rund 10 – 15 Minuten. Bitte beantworten Sie alle Fragen so gewissenhaft wie möglich. Wenn Sie sich nicht sicher bezüglich einer Antwort sind, schätzen Sie einfach so gut Sie können.

innoage

1. **Wie viele Tage pro Jahr sind Sie campen?**

    _____ Tage

2. **Seit wie vielen Jahren fahren Sie regelmäßig zum Camping?**

    _____ Jahre

3. **Wie würden Sie Ihre Campingexpertise einschätzen?**

    ☐ Sehr hoch

    ☐ Relativ hoch

    ☐ Weder hoch noch gering

    ☐ Relativ gering

    ☐ Sehr gering

4. **Wie interessiert sind Sie an Camping, verglichen mit anderen Campern?**

    ☐ Ich bin viel stärker interessiert

    ☐ Ich bin stärker interessiert

    ☐ Ich bin genauso interessiert

    ☐ Ich bin weniger interessiert

    ☐ Ich bin viel weniger interessiert

5. **Inwieweit stimmen Sie den folgenden Aussagen zu?**

	Trifft vollständig zu 1	2	3	4	Trifft überhaupt nicht zu 5
Ich weiß genau, welche Produkteigenschaften mir bei der Auswahl meiner Camping-ausrüstung wichtig sind.	☐	☐	☐	☐	☐
Ich nutze meine Ausrüstung intensiv.	☐	☐	☐	☐	☐
Ich habe einen guten Überblick über die am Markt verfügbare Ausrüstung.	☐	☐	☐	☐	☐
Ich kenne mich mit den Materialien und Einzelteilen meiner Ausrüstung aus.	☐	☐	☐	☐	☐

6. **Haben Sie jemals existierende (Camping-)Produkte verbessert oder hatten Sie Ideen für neue Produkte, die vorher nicht am Markt angeboten wurden?**

    Eine Produktidee/-verbesserung kann sich auf ein bereits bestehendes Produkt beziehen oder eine völlige Neuentwicklung sein.

    ☐ Ja *(weiter bei Frage 8)*   ☐ Nein *(weiter bei Frage 21)*

7. **In welchem Jahr haben Sie Ihre letzte Produktidee/-verbesserung entwickelt?**

    Eine Produktidee/-verbesserung kann sich auf ein bereits bestehendes Produkt beziehen oder eine völlige Neuentwicklung sein. Ebenso kann der Status der Innovation sehr unterschiedlich sein. Bitte denken Sie an Ihre letzte konkrete Innovation, egal ob es sich hierbei bisher nur um eine reine Idee handelt, oder bereits Skizzen, Modelle, Prototypen oder sogar ein fertiges Produkt existiert.

    _____

8. **Wie weit haben Sie Ihre Idee bislang entwickelt?**
   ☐ Ich habe eine mögliche Lösung im Kopf.
   ☐ Ich habe konzeptionelle Beschreibungen/Skizzen angefertigt.
   ☐ Ich habe einen Prototyp gebaut, der verlässlich genug ist, so dass ich ihn nutzen kann.
   ☐ Andere benutzen Prototypen, die auf meiner Idee basieren.
   ☐ Die Idee wurde bereits kommerzialisiert und ist im Handel verfügbar.

9. **Wie lange haben Sie benötigt, von der ursprünglichen Idee bis zum derzeitigen Entwicklungsstand?**
   Bitte geben Sie an, wie lange Sie sich mit der Entwicklung insgesamt beschäftigt haben (unterbrochen von anderen Tätigkeiten).
   ☐ < 1 Woche
   ☐ 1 – 4 Wochen
   ☐ 1 – 3 Monate
   ☐ 3 – 6 Monate
   ☐ 6 – 12 Monate
   ☐ > 12 Monate

10. **Wie häufig überarbeiten oder verbessern Sie Ihre Campingausrüstung selbst?**
    ☐ (Fast) alle meine Ausrüstungsgegenstände
    ☐ Meistens
    ☐ Gelegentlich
    ☐ Selten
    ☐ Nur dieses eine Mal

11. **Welche Teile Ihrer Campingausrüstung überarbeiten oder verbessern Sie normalerweise?**
    _____

12. **Haben Sie Ihre Produktidee/-verbesserung allein oder gemeinschaftlich mit anderen entwickelt?**
    ☐ Allein
    ☐ Gemeinschaftlich – Ich war die treibende Kraft
    ☐ Gemeinschaftlich – Alle hatten gleichen Anteil
    ☐ Gemeinschaftlich – Jemand anderes war die treibende Kraft

13. **Haben Sie für die Umsetzung Ihrer Produktidee/-verbesserung Hilfe von anderen benötigt?**
    ☐ Ja   ☐ Nein

14. **Wie würden Sie Ihre Produktidee/-verbesserung klassifizieren?**

    ☐ Neue Funktionalität

    ☐ Komfortverbesserung

    ☐ Kostenreduzierung

    ☐ Zeitersparnis

    ☐ Bessere Kompatibilität bzw. „Passgenauigkeit"

    ☐ Anderes: _____

15. **Bitte beschreiben Sie kurz Ihre Produktidee/-verbesserung.**
    Nutzen Sie ggf. die Rückseite, sollte der Platz nicht ausreichen.

    _____
    _____
    _____
    _____
    _____
    _____
    _____
    _____
    _____
    _____

16. **Bitte bewerten Sie Ihre Produktidee/-verbesserung bezüglich der folgenden Kriterien:**

16a. Neuheit:	Komplett neues Produkt	☐1	☐2	☐3	☐4	☐5	Kleine Verbesserung/ Geringe Modifikation
16b. Kreativität:	Sehr kreativ	☐1	☐2	☐3	☐4	☐5	Überhaupt nicht kreativ
16c. Technische Qualität:	Neue Technologie/ Hoher technischer Anspruch	☐1	☐2	☐3	☐4	☐5	Bekannte Technologie/ Geringer technischer Anspruch

17. **Angenommen, dass Ihre Produktidee/-verbesserung produziert würde, bitte bewerten Sie den Nutzen Ihrer Idee für Camper…**

17a. …heutzutage	Sehr hoch	☐1	☐2	☐3	☐4	☐5	Sehr gering
17b. …in der Zukunft	Sehr hoch	☐1	☐2	☐3	☐4	☐5	Sehr gering

18. **Angenommen, dass Ihre Produktidee/-verbesserung produziert und zum Verkauf angeboten wird, bitte schätzen Sie ein, wie viele Camper Ihre Idee…**

18a. …heutzutage kaufen würden	Viele	☐1	☐2	☐3	☐4	☐5	Wenige
18b. …in der Zukunft kaufen würden	Viele	☐1	☐2	☐3	☐4	☐5	Wenige

19. Inwieweit treffen die folgenden Aussagen auf Sie zu?

	Trifft vollständig zu 1	2	3	4	Trifft überhaupt nicht zu 5
Ich wollte mit Hilfe der Idee/Verbesserung Geld verdienen.	☐	☐	☐	☐	☐
Ich wurde für meine Idee/Verbesserung finanziell unterstützt.	☐	☐	☐	☐	☐
Ich wollte das Produkt selbst nutzen.	☐	☐	☐	☐	☐
Es war schön, Anerkennung zu bekommen.	☐	☐	☐	☐	☐
Es machte mir Spaß, meine Campingausrüstung zu verbessern.	☐	☐	☐	☐	☐

20. Inwieweit treffen die folgenden Aussagen auf Sie zu?

	Trifft vollständig zu 1	2	3	4	5	6	Trifft überhaupt nicht zu 7
Ich erfahre normalerweise von neuen Campingprodukten und -lösungen bevor es andere tun.	☐	☐	☐	☐	☐	☐	☐
Ich habe stark profitiert vom frühen Einsatz neuer Campingprodukte.	☐	☐	☐	☐	☐	☐	☐
Ich habe Prototypen neuer Campingprodukte für Hersteller getestet.	☐	☐	☐	☐	☐	☐	☐
Unter Campern werde ich als „Vorreiter" angesehen.	☐	☐	☐	☐	☐	☐	☐
Ich habe (neue) Bedürfnisse, welche durch existierende Campingprodukte nicht befriedigt werden.	☐	☐	☐	☐	☐	☐	☐
Ich bin unzufrieden mit der existierenden Campingausrüstung.	☐	☐	☐	☐	☐	☐	☐

21. Inwieweit treffen die folgenden Aussagen auf Sie zu?

	Trifft vollständig zu 1	2	3	4	Trifft überhaupt nicht zu 5
Ich kann meine eigene Ausrüstung reparieren.	☐	☐	☐	☐	☐
Ich kann anderen Campern helfen, Probleme mit ihrer Ausrüstung zu lösen.	☐	☐	☐	☐	☐
Ich bin handwerklich begabt und habe Spaß am basteln.	☐	☐	☐	☐	☐
Ich kann technische Änderungen an meiner Campingausrüstung selbst durchführen.	☐	☐	☐	☐	☐
Ich versuche bei meiner Ausrüstung immer auf dem neuesten Stand zu sein in Bezug auf Materialien, Neuheiten und Einsatzmöglichkeiten.	☐	☐	☐	☐	☐
Ich bin ein großer Fan von technischen Aspekten im Camping-/Caravaning-Bereich.	☐	☐	☐	☐	☐
Ich habe einen technischen Hintergrund in meinem Beruf/meiner Ausbildung.	☐	☐	☐	☐	☐

22. Wie alt sind Sie?

_____ Jahre

23. Die meisten Menschen scheinen nicht nur ihr chronologisches „Geburtsalter" zu besitzen sondern darüber hinaus auch „andere Alter". Die folgenden Fragen wurden entwickelt, um mehr über Ihr „inoffizielles" Alter herauszufinden. Bitte geben Sie pro Frage an, zu welcher Altersgruppe Sie sich zugehörig fühlen.

	20-24 Jahre	25-29 Jahre	30-34 Jahre	35-39 Jahre	40-44 Jahre	45-49 Jahre	50-54 Jahre	55-59 Jahre	60-64 Jahre	65-69 Jahre	70-74 Jahre	75-79 Jahre	80-84 Jahre	85-89 Jahre
Ich **FÜHLE** mich, als ich in meinen ... wäre.	☐	☐	☐	☐	☐	☐	☐	☐	☐	☐	☐	☐	☐	☐
Ich **SEHE SO AUS**, als ob ich in meinen ... wäre	☐	☐	☐	☐	☐	☐	☐	☐	☐	☐	☐	☐	☐	☐
Ich **VERHALTE** mich, als ob ich in meinen ... wäre.	☐	☐	☐	☐	☐	☐	☐	☐	☐	☐	☐	☐	☐	☐
Meine **INTERESSEN** sind vor allem die einer Person in seinen/ihren ...	☐	☐	☐	☐	☐	☐	☐	☐	☐	☐	☐	☐	☐	☐
ANDERE SCHÄTZEN MICH ein, als ob ich ... wäre.	☐	☐	☐	☐	☐	☐	☐	☐	☐	☐	☐	☐	☐	☐

24. Welchen Familienstand haben Sie?

    ☐ Ledig

    ☐ In einer Partnerschaft

    ☐ Verheiratet

    ☐ Geschieden

    ☐ Verwitwet

25. Wie hoch ist das gesamte monatliche Nettoeinkommen Ihres Haushalts?
    "Haushalt" bezieht sich auf das Gesamteinkommen von Ihnen und Ihrem Partner/Ihrer Partnerin.

    ☐ < 1.000 Euro / Monat

    ☐ 1.000 – 2.000 Euro / Monat

    ☐ 2.000 – 3.000 Euro / Monat

    ☐ 3.000 – 4.000 Euro / Monat

    ☐ 4.000 – 5.000 Euro / Monat

    ☐ > 5.000 Euro / Monat

    ☐ Keine Angabe

26. Welcher Anteil des gesamten monatlichen Nettoeinkommens Ihres Haushalts steht Ihnen zur freien Verfügung (d.h. ist normalerweise nicht bereits verplant)?

    _____ Prozent

27. Wie viel Zeit eines Durchschnittstages (von 8 – 23 Uhr) haben Sie für selbst gewählte Tätigkeiten zur freien Verfügung (d.h. ist normalerweise nicht bereits verplant)?

    _____ Stunden

28. Geben Sie bitte Ihr Geschlecht an.

    ☐ Weiblich   ☐ Männlich

29. Wo haben Sie bis 1990 gelebt?

    ☐ Gebiet der ehemaligen DDR   ☐ Alte Bundesländer der BRD   ☐ Anderes: _____

30. Welchen höchsten (Aus-)Bildungsabschluss haben Sie?

    ☐ Haupt-/Volksschulabschluss

    ☐ Mittlere Reife

    ☐ Fachhochschulreife/Abitur

    ☐ Abgeschlossene Ausbildung

    ☐ Universitätsabschluss

    ☐ Anderes: _____

31. Sind Sie derzeit berufstätig?
☐ Ja, Vollzeit
☐ Ja, Teilzeit
☐ Nein, derzeit arbeitssuchend
☐ Nein, bereits pensioniert

32. Welcher Berufsgruppe gehören Sie derzeit/gehörten Sie zuletzt an?
☐ Angehöriger der regulären Streitkräfte
☐ Führungskräfte
☐ Akademische Berufe
☐ Techniker und gleichrangige nichttechnischen Berufe
☐ Bürokräfte und verwandte Berufe
☐ Dienstleistungsberufe und Verkäufer
☐ Fachkräfte in Land- und Forstwirtschaft und Fischerei
☐ Handwerks- und verwandte Berufe
☐ Bediener von Anlagen und Maschinen und Montageberufe
☐ Hilfsarbeitskräfte
☐ Anderes: _____

33. Als was sind Sie/waren Sie zuletzt tätig?
☐ Angestellte/-r
☐ Arbeiter/-in
☐ Auszubildende/-r
☐ Selbstständige/-r
☐ Mithelfende/-r Familienangehörige/-r
☐ Beamter/Beamtin, Richter/-in
☐ Soldat/-in
☐ Nebenjobber/-in

34. Als was sind Sie/waren Sie zuletzt tätig?
Bitte geben Sie Ihre Berufsbezeichnung, inklusive Branche, an.

_____

Sie haben den Fragebogen nun vollständig bearbeitet. Wenn Sie möchten, können Sie hier noch Kommentare und Feedback zum Thema oder konkret zum Fragebogen hinterlassen.

_____
_____
_____
_____
_____
_____
_____

Vielen Dank für das Ausfüllen dieser Befragung!

Wenn Sie weitere Informationen über das Forschungsprojekt erhalten möchten, besuchen Sie unsere Homepage.

http://www.tuhh.de/innoage oder http://www.tuhh.de/tim

MIX
Papier aus verantwortungsvollen Quellen
Paper from responsible sources
FSC® C105338

If you have any concerns about our products,
you can contact us on
ProductSafety@springernature.com

In case Publisher is established outside the EU,
the EU authorized representative is:
**Springer Nature Customer Service Center GmbH
Europaplatz 3, 69115 Heidelberg, Germany**

Printed by Libri Plureos GmbH
in Hamburg, Germany